00672526

12515

GC
1015
F7
1971

Friedmann, Wolfgang
The future of the
oceans

Date Due

PLATTE COLLEGE RESOURCE CENTER
COLUMBUS, NEBRASKA

The Future of the Oceans

THE
FUTURE
OF THE
OCEANS

WOLFGANG FRIEDMANN

George Braziller
NEW YORK

Standard Book Number: 0-8076-0602-2, cloth
0-8076-0601-4, paper
Library of Congress Catalog Card Number: 70-142045
First Printing
Printed in the United States of America by Ray Freiman and Company
Designed by Selma Ordewer

Acknowledgments

The author wishes to thank his colleague, Professor Louis Henkin of Columbia University, for reading the manuscript and giving his valuable comments.

The author and publisher thank the following for permission to reprint certain materials included in this book from the books and periodicals listed below:

Accademia Nazionale dei Lincei—*The Symposium on the International Regime of the Seabed,* Rome, 1970:

Christy, F. T., Jr., "Marigenous Minerals: Wealth, Regimes and Factors of Decision."

Craven, J. P., *"Res Nullius de facto.* The Limits of Technology."

Evensen, J., "The Military Uses of the Deep Ocean Floor and its Subsoil—Present and Future."

Holt, S. J., "The Living Resources of the Seabed."

Muscarella, G., "International Regime of the Seabed; Oil and Natural Gas—Exploration and Exploitation of Petroleum in Deep Water."

Revelle, R., "Scientific Research on the Seabed."

British Yearbook of International Law, Vol. 27, on behalf of the Royal Institute of International Affairs. Lauterpacht, H., "Sovereignty over Submarine Areas" (London, 1950).

Center for the Study of Democratic Institutions— *Pacem in Maribus,* an Occasional Paper:

LaQue, F., "Deep Ocean Mining: Prospects and Anticipated Short-Term Benefits."

Stewart, C. and Pontecorvo, G., "Problems of Resource Exploitation: The Oil and Fishing Industries."

Columbia Journal of Transnational Law, Vol. 7. Oda,
S., "Proposals for Revising the Convention on the Continental Shelf" (New York, 1968).

Columbia University Press:

Andrassy, J., *International Law and the Resources of
the Sea* (New York, 1970).

Henkin, L., *Law for the Sea's Mineral Resources*
(New York, 1968).

Foreign Affairs—Brennan, G., "The Case for Missile Defense," April, 1969.

International and Comparative Law Quarterly—Jennings, R. Y., "The Limits of Continental Shelf Jurisdiction" (New York, 1969).

Law of the Sea Institute—"A Hypothetical Division of
the Sea Floor," a map reproduced, with modifications,
from the material prepared by Robert H. Warsing as
an annex to *The Law of the Sea: Proceedings of the
Second Annual Conference of the Law of the Sea Institute, 1967.*

New York Times Magazine—Galton, L., "Aquaculture
Is More than a Dream" (New York, 1967).

A. Pedone—Scelle, V., *Plateau Continental et Droit International* (Paris, 1955).

Scientific American, September, 1969:

Bascom, W., "Technology and the Ocean."

Holt, S. J., "The Food Resources of the Ocean."

Wenk, E., Jr., Map of Ocean-Floor Resources.

For

May

Contents

Preface

THIS is not a neutral book. The facts and data given are, I believe, beyond question, and I am greatly indebted for them to the many learned symposia and articles that have been published, particularly in the last three years, by specialists in the different areas of oceanbed exploration and administration.

But the views expressed here are no more neutral than those of the many institutions, groups, and individuals who are pleading for the partition of the oceans. Some have done so openly, others under the guise of legal "interpretations." To the present writer, a continuation of the trend of recent years can only spell unmitigated disaster. It can end only in military, political, and economic confrontation below as well as above the seas, while the dangers of further pollution of marine environment will be immeasurably increased by unregulated competition. Nor can a professor of international law be neutral toward the methodical undermining of the freedom of the seas, which is an inevitable consequence of the progressive appropriation of growing sections of the oceanbed by coastal nations. I happen to believe that the existing measure of international order in the world is totally insufficient to cope with the urgent tasks of civilized survival that will confront us with increasing urgency in the decades to come. The freedom of the seas cannot remain a *laissez-faire* freedom. In our overcrowded world, navigation, as well as the exploitation of the living and mineral resources of the sea, must be the subject of planning and regulation for the common benefit of mankind. But an unholy alliance of governments, interest groups, and lawyers is working toward nothing less than the partition of the seabed and therefore, inevitably, of the oceans themselves. To remain silent or passive toward such a development is to participate in the destruction of the principal achievements of international law in the last few centuries and to accept the preparation of an Orwellian nightmare.

The Nixon statement of May, 1970, and the subsequent United States Draft Convention offer some hope for putting a halt to the partition of the oceans and establishing a reasonable accommodation between the interests of the coastal states and the needs of the international community. But many obstacles remain to be overcome before even this moderate proposal can hope to be translated into international law.

The present book is a modest contribution to the discussion of one of our most urgent problems. An appraisal of the major issues posed by the growing exploration and exploitation of the oceanbed, and their consequences for the freedom of the seas, is followed by an attempt to make a few meaningful proposals for the creation of an authoritative oceanbed control and various forms of international ocean enterprise that could channel the present destructive race into constructive and worldwide co-operation.

W. F.

New York, December 1970

1

The Challenge

AMONG the many challenges that face mankind in the remaining decades of the twentieth century two stand out as of crucial importance for the very survival of civilization. One is the ecological problem of man's ability to cope with an environment of his own creation, which now threatens to overwhelm him. The other is the political problem of choosing between a competitive race of nations for power and wealth—a race that can only lead to the ultimate confrontation of a few superpowers—and ordered co-operation, in which countries can combine their purposes, their ingenuity, and their resources in an international order that envisages mankind as a whole.

The ecological and political challenges are closely connected. And of all the many areas in which a fateful choice must be made, there is none more important or urgent, both qualitatively and quantitatively, than the future status of the oceans.

From the dawn of history until a quarter of a century ago the seas have served two main purposes: communication and fishing. The age of discoveries, notably the explorations of the sixteenth and seventeenth centuries, opened up communications between continents that until then had remained unknown to each other. America, Asia, and in the eighteenth century, Australia, were discovered and penetrated by European explorers. Ships crossing the oceans began to link the continents, and the advent of the steamer intensified these links. The steamship also contributed to the development of new areas and mobilized the means of fishing. But these were quantitative rather than qualitative changes; it is only in the last few decades that the advent of whale and fish factory ships, and the introduction of new, intensified methods of killing fish have begun to create basic conservation and ecological problems.

Throughout the ages the deeper reaches of the oceans and the bottom of the seas remained hidden, an area of mysteries and legends, of sea monsters and spirits, but outside the practical concerns of mankind. The Truman Proclamation of September 28, 1945, marks the beginning of a new era, with the conversion of seven-tenths of the earth's surface into an area that before the end of the century may be as open to exploitation, transport, cultivation, and settlement on the oceanbed, as the three-tenths covered by land. From that date we can trace both the technological revolution that is opening up the oceanbeds to exploitation of both mineral and biological resources at an ever-increasing pace, and a new phase in international relations, in which the oceans, hitherto free to all for navigation and fishing, are subjected more and more to exclusive and competing national controls, and widening portions of the oceanbed—and therefore indirectly of the oceans themselves—are partitioned between rival states or groups of states.

The seas have often been the scene of naval battles, and some nations, notably Britain from the beginning of the seventeenth century to the beginning of World War II, derived their power mainly from naval supremacy. But never—since the *Mare Liberum* (1609) doctrine of Hugo Grotius prevailed over that of John Selden (*Mare Clausum*, 1635)—did the seven seas become closed or divided between nations in times of peace. The freedom of the seas—limited only by exclusive national territorial water zones ranging from a minimum of three miles to a maximum of twelve miles[1] with international rights of passage for peaceful purposes—was indeed the single most important accomplishment of the law of nations as it has developed over the last three and a half centuries. Today this freedom is not yet directly challenged, but as the oceanbeds are increasingly opened up at ever-greater depth to exploration and exploitation by various national and corporate interests, the limitations of the traditional freedoms of the sea become both more numerous and more disturbing. Exclusive national claims to exploit the resources of steadily widening areas of the oceanbed inevitably lead to political assertions designed to buttress such claims. And as drilling rigs, floating islands, stationary platforms, submersibles and artificial structures above and below the surface of the sea multiply, the traditional freedoms of fishing and shipping, however strongly they may be affirmed theoretically, must be qualified, restricted, and ultimately excluded.

The Truman Proclamation is generally taken to be the starting point of a new era in the role of the oceanbed. Because of its legal status, its economic significance, and its potential for political and military conflict it triggered a series of rapidly expanding and accelerating national claims to exclusive control over the conti-

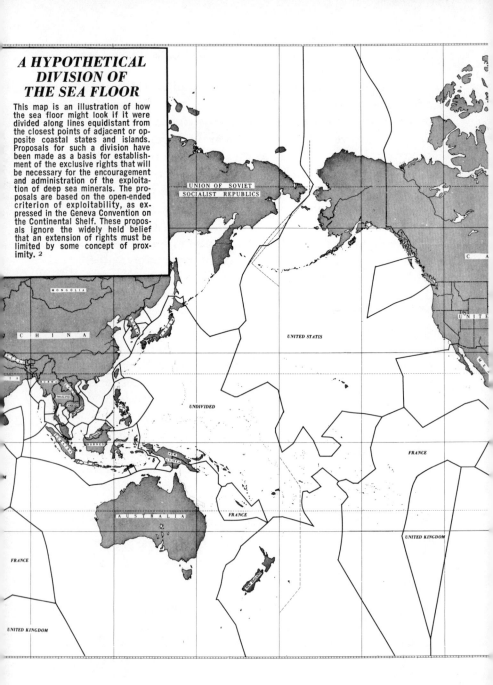

A HYPOTHETICAL DIVISION OF THE SEA FLOOR

This map is an illustration of how the sea floor might look if it were divided along lines equidistant from the closest points of adjacent or opposite coastal states and islands. Proposals for such a division have been made as a basis for establishment of the exclusive rights that will be necessary for the encouragement and administration of the exploitation of deep sea minerals. The proposals are based on the open-ended criterion of exploitability, as expressed in the Geneva Convention on the Continental Shelf. These proposals ignore the widely held belief that an extension of rights must be limited by some concept of proximity. 2

UNION OF SOVIET SOCIALIST REPUBLICS

MONGOLIA

C H I N A

THAILAND

BORNEO

A U S T R A L I A

NEW ZEALAND

UNITED STATES

UNDIVIDED

FRANCE

FRANCE

FRANCE

UNITED KINGDOM

UNITED KINGDOM

GREENLAND

ICELAND

DENMARK

UNION OF SOVIET S

BOUNDARIES

Basic map is taken from
H.O. 1262F.
_ _ _ _ International Boundary
——— Division of the
sea floor
..........other separations and
1937 boundaries

Scale 1:39,000,000 at Equator

CANADA

ENGLAND

GERMANY

POLAND

FRANCE

RUMANIA

TURKEY

SPAIN

IRAN

UNITED STATES

PORTUGAL

SPAIN

ALGERIA

LIBYA

UNITED ARAB REPUBLIC

SAUDI ARABIA

UNITED KINGDOM

MAURITANIA

NIGER

CHAD

SUDAN

FRANCE

BARBADOS

NET

GUINEA

IVORY COAST

GHANA

CENTRAL AFRICAN REPUBLIC

ETHIOPIA

VENEZUELA

COLOMBIA

DEMOCRATIC REPUBLIC OF THE CONGO

TANZANIA

UNITED KINGD

ECUADOR

BRAZIL

ANGOLA

ZAMBIA

RHODESIA

FR

BOLIVIA

SOUTH WEST AFRICA

BECHUANALAND

FR

REPUBLIC OF SOUTH AFRICA

FRANCE

UNITED KINGDOM

NORWAY

UNI

ARGENT

nental shelf, defined in the Proclamation as an "extension of the land mass of the coastal nation and thus naturally pertinent to it. . . ." Ever since the Proclamation nation after nation has claimed exclusive sovereignty over the continental shelf, an international convention on the subject has been concluded, and many states that refused to sign the convention, or that have no exploitable continental shelf, have staked compensatory claims to portions of the sea. The concept of the continental shelf itself has proved to be the starting point for constantly expanding claims to large sectors of the seas, which may conceivably end in the apportionment of the oceans by various maritime nations.

The Truman Proclamation is of such importance in the history of international relations that it merits quotation in full:

WHEREAS the Government of the United States of America, aware of the long range world-wide need for new sources of petroleum and other minerals, holds the view that efforts to discover and make available new supplies of these resources should be encouraged; and

WHEREAS its competent experts are of the opinion that such resources underlie many parts of the continental shelf off the coasts of the United States of America, and that with modern technological progress their utilization is already practicable or will become so at an early date; and

WHEREAS recognized jurisdiction over these resources is required in the interest of their conservation and prudent utilization when and as development is undertaken; and

WHEREAS it is the view of the Government of the United States that the exercise of jurisdiction over the natural resources of the subsoil and sea bed of the continental shelf by the contiguous nation is reasonable and just, since the effectiveness of measures to utilize or conserve these resources would be contingent upon cooperation and protection from the shore, since the continental shelf may be regarded as an extension of the land-mass of the coastal nation and thus naturally appurtenant to it, since these resources frequently form a seaward extension of a pool or deposit lying within the territory, and since self-protection compels the coastal nation to keep close watch over activities off its shores which are of the nature necessary for utilization of these resources;

NOW, THEREFORE, I, HARRY S. TRUMAN, President of the United States of America, do hereby proclaim the following policy of the United States of America with respect to the natural resources of the subsoil and sea bed of the continental shelf.

Having concern for the urgency of conserving and prudently utilizing its natural resources, the Government of the United States regards the natural resources of the subsoil and sea bed of the continental shelf beneath the high seas but contiguous to the coasts of the United States as appertaining to the United States, subject to its jurisdiction and control. In cases where the continental shelf extends to the shores of another State, or is shared with an adjacent State, the boundary shall be determined by the United States and the State concerned in accordance with equitable principles. The character as high seas of the waters above the continental shelf and the right to their free and unimpeded navigation are in no way thus affected.

Herein lie all the factors that account for the ever-quickening race to the bottom of the ocean: the need for new sources of petroleum and other minerals; the technological breakthrough that makes utilization of the seabed mineral resources increasingly practical; the claim to exclusive national jurisdiction "in the interest of . . . conservation and prudent utilization. . ."; and the definition of the continental shelf as an extension of the territory of the coastal nation and thus subject to its sovereignty. The subsequent twenty-five years have

seen a broadening and intensification of these factors
and their related claims, to an extent exceeding even the
most sanguine expectations of the framers of the Proclamation. The technology of oceanbed exploration and
exploitation is proceeding at a rate that makes the scientific data of even a few years ago obsolete. And the areas
to which the various coastal states are granting exploration licenses is expanding from year to year, both in
depth and width.

As a direct result of these developments, international
maritime law has become the subject of a heated controversy that reflects the battle between nationalists
and internationalists. The former follow the long-
standing legal tradition of legitimizing the claims of
their governments, however extravagant they may be.
The latter are fighting an increasingly intensive battle
for the creation of a new co-operative international
ocean regime to serve the interests of small as well as
large states, of the undeveloped as well as the technologically advanced, and of the landlocked as well as the
maritime nations. It will be the purpose of this short
book to portray for the interested and concerned
nonspecialist the overwhelming importance as well as
the complexity of this challenge, and the interdepend-
ence of the scientific, technological, economic, and legal
aspects of the matter.

---

[1] The mile referred to throughout this book is the nautical or marine mile, which
is 800 feet longer than the land mile. The meter used as a unit of measure in the
following chapters equals 3.28 feet.
[2] F. T. Christy, Jr., and H. Herfindahl in a note to the map, "A Hypothetical
Division of the Sea Floor," prepared for the Law of the Sea Institute, 1967.

2

Some Basic Concepts and Data

THE gradual conversion of the oceanbed which constitutes the major surface of our globe, from a mystery unknown to man beyond the shallow depths in which he fished, to an explorable and exploitable area with a topography of mountains, ridges, slopes, plains, trenches, and abysses, makes it necessary for the nonspecialist to be familiar with certain basic concepts and categories formerly known only to the geologist and the marine biologist. For within the life span of our generation the various layers and structures of the oceanbed will become as important to the life and activities of nations as the mountains, valleys, and rivers that have determined the geographical and political topography of the continents up until now.

We will first briefly define the various layers of the oceanbed as they move outward from the shore to the abyssal depths of the seas.

First in line is the *continental shelf*. As defined by a group of geologists assembled in 1969 in Rome for a

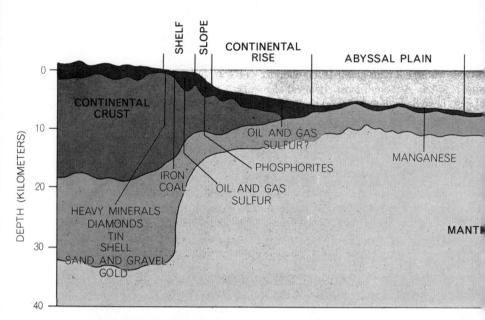

OCEAN-FLOOR RESOURCES that are known or believed to exist in the various physiographic provinces are indicated on a schematic cross section of a generalized ocean basin extending from a continent out to a mid-ocean ridge. Some of these resources are now being exploited but others

symposium[1] on the International Regime of the Seabed it consists of "the zone around the continent extending from the low water line to the depth at which there is usually a marked increase of declivity to greater depth." The shelf ends at the point where this marked increase occurs. The waters of this so-called "shelf edge" range in depth from fewer than 60 meters to more than 500 meters, and average about 130 meters. Simultaneously with this claim to exclusive jurisdiction over the subsoil and seabed resources of the United States' continental shelf in the Proclamation, Truman defined the continental shelf in a White House press release as "submerged land which is contiguous to the continent and which is covered by no more than one hundred fathoms (600 ft.) of water. . . ." Thirteen years later, Article I

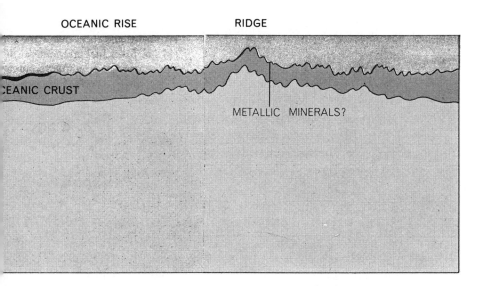

OCEANIC RISE   RIDGE

EANIC CRUST

METALLIC MINERALS?

may not be economic for years. Sedimentary layers (*black*) are the most likely site of recoverable raw materials. The chart (*below*) shows what percent of the ocean floor's 140 million square miles of area is occupied by each province.

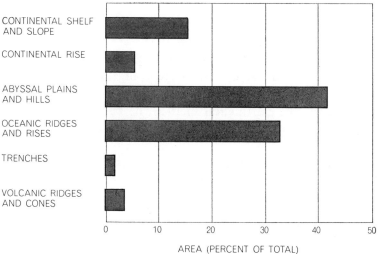

CONTINENTAL SHELF AND SLOPE

CONTINENTAL RISE

ABYSSAL PLAINS AND HILLS

OCEANIC RIDGES AND RISES

TRENCHES

VOLCANIC RIDGES AND CONES

0   10   20   30   40   50

AREA (PERCENT OF TOTAL)

of the Geneva Convention on the Continental Shelf accepted the roughly equivalent criterion of "a depth of 200 metres," although, in a disastrous clause that opened the door for an accelerating race for exclusive jurisdiction to much greater depths, it added the criterion of *exploitability* beyond the depth of 200 meters as a justification of extending continental-shelf jurisdiction. The continental shelves underlie only 7.5 per cent of the oceans, but they equal 18 per cent of the earth's total land area.

The seaward extension of the continental shelf is called the *continental slope,* a zone that extends from the shelf edge to a depth of 1,200-3,500 meters. The outer edge of the continental slope generally marks the boundary between the low-density rocks of the continent and the high-density rocks of the deep ocean floor.

The *continental rise* is a zone that borders the base of many, although not all, continental slopes. With a generally smooth declivity it goes to a depth of 3,500 to 5,500 meters.

The continental shelf, continental slope, and continental rise together are known as the *continental margin.* As we shall see, there is increasing pressure, particularly on the part of the oil interests of the technologically advanced nations, to substitute the continental margin for the continental shelf as the zone of national jurisdiction and sovereignty over seabed and subsoil resources.

Beyond the continental margin lies the *abyssal plain* —the extremely flat area of the deep ocean floor, at depths generally in excess of 5,000 meters.

Lastly, seamounts, atolls, and shallow banks, generally of volcanic origin and with diameters between 12 meters and 150 kilometers are scattered throughout the oceans. According to Henry Menard,[2] 10,000 volcanoes with a relief of more than one kilometer above the sea floor exist in the Pacific Ocean alone, the most famous of which is Bikini. Many others have been found in the Coral Sea, the South China Sea, the Caribbean, and the northeast Atlantic. These formations are of potential political and legal importance because they rise from the deep, outside the continental-shelves.

National States and the Continental Margin

If the ultimate control of the exploitation and distribution of the resources of land and sea were in the hands of the United Nations or some other supranational authority, the many inequalities in the existence, extension, and exploitability of the continental shelves —and the wider continental margins—would not greatly matter. The combined resources of the seabed could form an international pool and be distributed among nations according to some agreed equitable principles. These would be based presumably on such factors as density of population, gross national product, richness in land-based resources, and the state of economic development. Under such a worldwide regime and system of distribution the resources of the oceanbed could in fact be used to mitigate the many inequalities that today make different states rich or poor in natural resources. However, a world federation is a distant dream, and the United Nations has so far failed to develop into an effective supranational authority, in the military, political, or economic sense.

After World War II Sir John Boyd-Orr, a British agricultural expert, envisaged a world food authority that would redistribute vital foodstuffs from surplus to deficient areas. This plan was defeated, giving place to the Food and Agricultural Organization, a United Nations specialized agency with purely advisory and informational functions.

Not only has the world remained divided into sovereign nation states, but the number of these states has more than doubled in the postwar period and now stands at well over 130. They range in population from the 700 million to 800 million of mainland China to the 80,000 of the Maldive Islands and include the United States and the Soviet Union as well as dozens of states whose economic product would not equal that of some single corporations in the United States. The multiplication of "sovereign" nation-states represents one of the supreme and tragic ironies of our time. It has occurred in an era when distances have shrunk to a fraction of their former significance, when jet planes cross the Atlantic in a few hours and missiles in a few minutes, when radio, television, and satellites in outer space have engendered a total revolution in communications, and when sovereignty—for the vast majority of today's nations the symbol and focus of emotional loyalties—has become a fiction in terms of defense, political independence, or economic viability. Because the political organization of our contemporary world is still based on the concept of national legal supremacy, the extension of sovereignty to submarine areas adjacent to the coast (to comprise at a minimum the continental shelf and at a maximum the continental margin) has in most cases magnified the existing inequalities between nations for two geographical reasons.

First, there are at the present time twenty-nine landlocked states, which have no coast whatsoever and therefore cannot claim any adjacent submarine area for their exclusive jurisdiction. While a few of these states —such as Austria and Switzerland—are among the world's industrially developed countries, the great majority of the landlocked countries is found among the new and poorer states of Africa, with a few in Asia, Latin America, and Eastern Europe. Moreover, as Dr. Francis T. Christy, of Resources for the Future, has observed, about fifty other nations "face small seas, have little more than toeholds on the oceans, or have vistas broken by foreign islands off their coasts. Of the remaining 50, about half have small to moderate coastlines and the rest, extensive shores, on the open oceans."[3] In fact, only about two dozen nations would gain substantially from very wide limits of national oceanbed jurisdiction. But since the extension of sovereignty still has a magic appeal to today's nation-states—old and new, developed and undeveloped—each and every state will seek to extend the geographical area of its sovereignty, regardless of how small the benefits from such an extension might be.

Second, not all maritime nations have continental shelves, as defined earlier. Certain coastlines—notably those of part of the west coast of North America and considerable portions of the west coast of South America—decline steeply. Since the shelf is the area in which the greatest number of fish and exploitable minerals are concentrated, this means that certain states are excluded from the sudden accretion of national wealth that has come with the extension of national sovereignty to their resources. This sense of inequality has led a steadily growing number of Latin American states to proclaim a territorial water zone of 200 miles under the complete sovereignty of the coastal state. Compared with the usual territorial water limits of from three to twelve miles, this would seem to be something of an overcompensation for what nature neglected.

The fact that certain nations are far better equipped

to exploit oceanbed resources than others adds a third inequality, not on geographical but on technological grounds. Although, as we shall see, certain legal arrangements could mitigate this technological handicap, it gives an initial advantage to the few countries that combine exploitable seabed and subsoil resources with advanced technology. The undeveloped countries still make up a large majority of the world's nations, and although individually and collectively they have asserted their claim to the exploitation of natural resources over which they exercise sovereign rights and are gradually building up a reservoir of trained scientists, engineers, and administrators, the gap remains wide and will continue to be for many years. The various aspects of ocean technology require an even greater scientific and technological sophistication as well as capital investment than most land-based operations.

In short, the extension of sovereign rights to both mineral and living resources of the continental margin has tended to accentuate rather than mitigate existing national inequalities, creating a new potential for political and social tension, and rendering even more important the need for some form of international oceanbed control for the redistribution of resources.

---

[1] A collection of essays prepared for this symposium was published by the Accademia Nazionale dei Lincei under the title *The Symposium on the International Regime of the Seabed*. Hereafter, all references will be to *Rome Symposium*. (See Bibliography, p. 126)

[2] Quoted by A. J. Guilcher, "The Configuration of the Ocean Floor and Its Subsoil: Geopolitical Implications," *Rome Symposium*, p. 25.

[3] Francis T. Christy, Jr., "Marigenous Minerals: Wealth, Regimes and Factors of Decision," *Rome Symposium*, p. 146.

3

Ocean Bed Resources —
A Summary Survey

THE economic importance of ocean resources in
the decades to come is determined basically by two criteria: on the one hand, the kind of resources actually or
potentially available for exploitation; on the other
hand, their accessibility, determined by such factors
as the depth at which the resource occurs in exploitable
quantities and the geological structure of the seabed
wherein it is found. The importance of depth and geological structure is, of course, closely related to the
available methods of exploitation. If, for example, it is
either technically impossible or economically prohibi-
tive to drill for oil, natural gas, or other minerals at a
depth exceeding 50 meters, the occurrence of rich depos-
its at depths of 100 meters or 600 meters is of purely theoretical interest. Even during the thirteen years that have
elapsed since the Continental Shelf Convention of 1958,
the technology of oceanbed exploration, seabed stations,
oceanbed drilling at ever-increasing depths, and transportation along ocean floors have advanced at such a

dramatic rate that the importance of oceanbed resources has expanded dramatically. Ships now exist that can go up, down, sideways, and can flip. They can swim, fly, and dive, and they are able to navigate more precisely than ever before. These refinements, coupled with the ability to examine the ocean bottom in detail from the surface by means of television and side-looking sonar, have produced remarkable results.[1] Consequently, the legal status of the oceanbed has become a matter of crucial importance. In the words of a famous French proverb, *L'appétit vient en mangeant* ("Appetite comes with eating"). If, a quarter of a century ago, control of the seabed and subsoil resources of the continental shelf to a depth of 200 meters seemed more than adequate to satisfy the needs of the foreseeable future, today rights over the continental margin—extending from 3,000 meters to 4,000 meters deep—have become sufficiently desirable to be the subject of expanding national claims and bitter legal controversy.

Exploitable marine resources can be divided into three general categories. The first and only one that has been the subject of economic exploitation until very recently consists of fish and other free-moving organisms.[2]

The present ocean harvest was recently estimated at about 55 million metric tons per year, representing an income of approximately $8 billion. Ninety per cent of this consists of fin fish, the rest of whales, crustaceans, and mollusks, as well as other invertebrates. Three-quarters of this total harvest is taken by fourteen countries. What is perhaps even more significant is the rate of increase in the ocean harvest. In the century from 1850 to 1950 the world catch increased tenfold, at an average rate of about 25 per cent each decade. It doubled in the subsequent decade. Since 1938 the world fish catch has more than tripled. The Food and Agricultural Organization of the United Nations (FAO) has estimated the total 1968 catch at 64 million metric tons, about eleven-twelfths consisting of marine fishes. Of this total harvest half is consumed by humans, while the rest becomes livestock feed.[3]

The habits of such creatures as whales, salmon, and eel are still largely unexplored, but obviously these animals, together with the other pelagic fishes, move across the oceans in total disregard of national boundaries, making crucial the question of the extent of the freedom of the oceans. Together with freedom of navigation, unrestricted fishing—except for very narrow territorial waters—has been the most important aspect of the doctrine of the freedom of the seas. The growing tendency of states to extend their exclusive national fishery zones—in some cases as far as 200 miles from the coast—is one of the many grave encroachments upon this freedom.

Second, there are organisms, such as oysters, which are either stationary or move within very limited distances, and whose extraction has therefore been compared to "a type of agricultural processing."[4]

Third, and of increasing importance, are the offshore minerals, of which oil and gas are by far the most important at the present time. It was the rapidly expanding consumption of oil and gas, coupled with the growing feasibility of offshore drilling, that led to the Truman Proclamation and created the rapid spread of claims for exclusive exploitation of the continental shelf throughout the world. The figures are certainly staggering. An Italian expert has estimated that the world's consumption of oil—which increases at a rate of one billion barrels per year—will have risen in twenty years to five times the present volume.[5] At the end of 1968 proved oil reserves were estimated at more than 65 billion tons. The explosive acceleration in the demand for oil and gas products accounts for the dramatic increase in offshore production. In 1966 offshore oil and gas production in the United States, which ten years earlier represented 1 per cent of the total, had risen to 10 per cent, and by 1969, to 15 per cent (*Wall Street Journal,* September 30, 1969).

On a worldwide basis, offshore oil production in 1966 was 6,000,000 barrels per day, roughly 15 per cent of the total production. Offshore oil reserves have recently been estimated as 20 per cent of the world's total reserves (*New York Times,* February 12, 1967). While the rate of progress in offshore exploitation of oil and gas will depend on a number of variables and constantly shifting equations between the relative merits of urgency, economic efficiency, political stability, legal security, and other relevant factors, pressure for expanding offshore production is certain to lead to a continuing expansion of this form of mining.

Compared with oil and natural gas, the present value of other seabed minerals is at this time relatively minor, except for sulphur, which is already mined in considerable quantities from the subsoil off the coast of Louisiana. In his historic memorandum of August, 1967, to the First Committee to the United Nations General Assembly, Ambassador Arvid Pardo of Malta gave a survey of the mineral resources of the seabed. He observed that the nodules, which are the principal form of seabed surface deposit, contain the following:

43 billion tons of aluminum, equivalent to reserves for 20,000

years at the 1960 world rate of consumption, as compared to known land reserves for 100 years;

358 billion tons of manganese, equivalent to reserves for 400,000 years, as compared to known land reserves of only 100 years;

7.9 billion tons of copper, equivalent to reserves for 6,000 years, as compared to only 40 years for land;

nearly one billion tons of zirconium, equivalent to reserves for 100,000 years, as compared to 100 years on land;

14.7 billion tons of nickel, equivalent to reserves for 150,000 years, as compared to 100 years on land;

5.2 billion tons of cobalt, equivalent to reserves for 200,000 years, as compared to land reserves for 40 years only;

three-quarters of a billion tons of molybdenum, equivalent to reserves for 30,000 years, as compared to 500 years on land.

In addition, the Pacific Ocean nodules contain 207 billion tons of iron, nearly 10 billion tons of titanium, 25 billion tons of magnesium, 1.3 billion tons of lead, 800 million tons of vanadium, and so on.

But many of these valuable minerals are in the deeper reaches of the oceanbed, so dispersed that their economic exploitation is not practicable in the foreseeable future. To give just one example,[6] it has been estimated that there are about 5,000,000 tons of gold in the oceans, which would represent $5 trillion at current prices, in theory a staggering value. But the cost of exploitation would be so great that this potential source of wealth is purely theoretical. And a leading metallurgist and former vice president of the International Nickel Company has observed that current activity in the recovery of metals from deep-ocean nodules—particularly manganese—is essentially exploratory and speculative, with no likelihood of any exploitation of deep-ocean nodules on a commercial scale before 1985.[7] Other expert estimates appear to concur that "commercial development of deep-sea minerals is not likely to be significant for many years to come. . . . After development becomes significant, it will still be many years before royalty or rental payments will become large."[8] In an estimate of the profitability of deep-sea manganese exploitation (the most important of the mineral nodules) made in 1969 by the Commission on Marine Science, Engineering and Resources, the conclusion was that even on the most favorable technological assumptions the return on capital invested in deep-sea manganese exploitation at best would be marginal. But this may not necessarily apply to all regions of the ocean. Thus the leading study on the subject[9] has estimated the average concentration of nickel nodules for the whole Pacific at .35 per cent; but samples taken from the tip of Baja California have shown nickel contents of 1.19 per cent to 1.46 per cent.[10] Considering the fantastic rate of technological progress in oceanbed technology, deep-sea mineral activity and the legal status of the oceanbed may well become a matter of increasing importance. The economic consequences of exploiting such minerals as copper or nickel in marketable quantities from nodule concentrations is potentially formidable. By making certain relatively scarce materials abundant, it could completely upset the international commodities market. This in turn could deeply affect the attitude of certain major producers, like Chile, and to a lesser extent Peru, which would have an economic interest in preventing exploitation of copper from the ocean floor. Both countries have led Latin American claims to a 200-mile territorial water zone, in which they could hinder exploitation, and in the deep seas beyond national jurisdiction they would certainly do everything in their power to prevent exploitation of such resources under the aegis of an international seabed authority. The mere possibility of future exploitation of these riches is already influencing the legal attitudes of different states.

However, in the foreseeable future, the continental shelf proper—and to a lesser but increasingly important extent, the wider continental margin—will absorb the greatest part of each nation's offshore activities, since 90 per cent of the world's marine food resources, now extracted at the rate of $8 billion per year, and nearly a fifth of the total world production of petroleum and natural gas, representing a value of about $4 billion, come from the continental shelves.[11] At the present time, the deepest offshore production operation is at a depth of 103 meters (in the Gulf of Mexico). This is well within the universally recognized minimum depth of the continental shelf, which is 183 meters. But the rate of expansion in depth exploration is extraordinary. As Stewart and Pontecorvo have observed, although it required nearly twenty years for oil companies to move from the 15-meter level to the 195-meter level of exploratory drillings, only one year later experimental drillings had gone to 366 meters. Some believe that there will be drilling at 2,000 meters by the late seventies. There is, however, a wide gap between exploration and exploitation. The cost of oil exploitation is staggering and requires investments of which only major states or giant corporations or consortia are capable. According to the *Wall Street Journal* (September 30, 1969), the cost of drilling 12,500 holes under United States waters and bringing the commercially profitable borings into production was about $13 billion, including rental and royalty payments, and the costs are constantly going up.

It remains to say a few words about the prevailing

techniques of offshore drilling, for these, like the eco-
nomics of the matter, have a profound impact on the
political and legal future of the oceans. At present the
most predominant method of exploiting offshore fields
is from a fixed platform, supported by piles fixed in the
seabed. Wells are drilled into the ocean floor through
conductor pipes, and oil or gas is pumped through a sea
line from the platform to the shore. But although it is
still feasible to install fixed platforms and connecting
sea lines up to depths of 200 meters (the minimum limit
of the continental shelf), the installation of such platforms will become increasingly uneconomical as operations proceed to greater depths. While the precise limits
of economic rentability for the fixed platform are still
under discussion, the experts foresee the increasing importance of alternatives to drilling from fixed structures. One alternative is the floating unit, which may be
a drilling vessel or a semisubmersible platform. Floating
vessels are specially built, and self-propelled ships move
from one spot to another supporting the drilling rig.
Semisubmersible platforms are structures made of large
columns buoyed underneath by caissons, with decks
above on which drilling rigs, living quarters, and equipment are installed; these floating platforms are towed to
different areas of exploitation, where they are partially
sunk and moored.

Another important development is the underwater
well, for which the wellhead is actually located on the
seabed. Several types of underwater wellheads have already been installed in depths of up to 100 meters, and
it is anticipated that before long it will be possible to
control them remotely with acoustic signals, completely
doing away with the intervention of divers.

All this goes together with rapid progress in underwater transportation and the possibility of stationing human beings on the ocean floor for prolonged periods. Jacques-Yves Cousteau and Albert Piccard are among the pioneers in developing different forms of submersibles that can be used to explore the deep seas and the seabeds. Among the most famous experiments are the American Sealab project laboratories, in which teams of ten men lived in steel chambers for forty-five days; Cousteau's successful attempts in 1965 to have a team of six oceanonauts live for twenty-three days in a chamber from which they dived to depths of 100 meters to 120 meters without support from a surface vessel; and the Soviet experiment in the Black Sea, where scientists spent ten days in a laboratory 57 feet below the surface.[12]

We may now sum up this very cursory survey of the astounding technological developments of the last decade. First, the economic pressure for the exploitation of oil and gas at ever-increasing depths, principally concentrated in the continental shelves, is an inevitable consequence of the constant acceleration in mankind's demand for oil, gas, and other vital minerals. Second, technological progress makes full exploitation of the resources of the continental shelf proper, and more selective exploitation of mineral resources in the continental margin beyond the shelf a practical prospect within the next decade. The exploration and exploitation of the seabed resources of the oceanbed beyond the continental margin remain a more distant prospect. Third, there has been amazing progress in man's ability to spend prolonged periods on or near the bottom of the ocean at depths well exceeding the limits of the continental shelf proper. The variety of technological devices for the exploration of offshore resources is constantly increasing and will result, for one thing, in the gradual replacement of the presently prevailing fixed platform on the surface of the water, by floating platforms, drilling vessels, submerged installations, and transportation devices along the bottom of the oceans.

Oceanbed Mining and Its Effects on Navigation and Fishing

The steady horizontal and vertical extension of mining operations, with the multiplication of fixed and floating platforms, drilling vessels, submerged platforms, submarine stations, maintenance service and diving equipment, feeder lines, offshore storage, and loading facilities, will increasingly curtail two of the most vital areas of the freedom of the seas—navigation and fishing. In the Gulf of Mexico, one of the most closely mined offshore areas, oil rigs have become so numerous that it has been necessary to provide "fairlanes" for shipping. In spite of the safety precautions—warning signals and markers that may be installed on the high seas and the continental shelf—in accordance with the Geneva Convention, which purports to safeguard the freedom of shipping and fishing, these traditional freedoms will soon be converted from primary rights into secondary licenses by the escalation of mining operations. Ships will be forced to pick their way through the water above the continental shelf, as they used to, with the aid of local pilots, through dangerous straits and uncharted local waters.

The plight of fishing is perhaps even more obvious. The effects of mining, drilling, and trawling in the continental margin have been summed up by Dr. Sidney J.

Holt of the Food and Agricultural Organization: First, the seabed may become locally polluted with solid spoil, which will change the quantity and quality of seabed life, thus indirectly affecting the animals that feed on it. Secondly, and more important, oil spillages —such as those which provoked the recent disasters off Santa Barbara and the coast of Nova Scotia—are bound to occur with increasing frequency in proportion to the multiplication of oil drillings, affecting the living organisms of the sea and their food supplies. Third, pollution at the surface—the result of spillage from oil tankers and leaks in oil rigs—can be transferred to the seabed by the sinking of the oil masses. Fourth, the physical disturbances caused by fishing may affect the whole seabed life. Fifthly, installations such as cables and pipe lines can interfere with trawling and vice-versa.[13]

In addition to the disturbances caused by expanding mining operations in the sea, there is another and older danger to the living resources of the ocean: overfishing by trawler fleets like those of the Soviet Union, Japan, and Norway, which include fish-processing ships and apply increasingly mechanized and indiscriminate methods of fishing. This kind of unregulated and mechanized competition has already led to the virtual extinction of many species of whales and the rapid dwindling of such sea staples as California sardines, Northwest Pacific salmon, and Barents Sea cod. It is the result of lack of adequate, internationally agreed conservation measures, of the ruthless application of modern technology to an ancient occupation that did not and could not disturb the ecology of marine life as it was practiced over the ages.

One conclusion emerges clearly from all that has been

said: as the world population swells at an exponential rate from the present 3.5 billion to what will be more than double that number by the end of the century, accompanied by progressive industrialization and mechanization of production, there will be a proportionate increase in the demand for the enormous food and mineral resources contained in the oceans. Gradually, the seas will be farmed, mined, and exploited, like the land. Let us not be taken unawares in this still relatively virgin area, as we were in the unregulated development of the land, for which we are now painfully suffering the consequences. There is no alternative to conscious and planned regulation of life on the seas—the surface as well as on the oceanbed. The basic choice facing mankind is whether exploitation will occur by a ruthless partitioning between competing nations or whether the increasingly important resources of the sea will be considered, in the formulation first referred to by President Johnson in 1966, as the "common heritage of mankind." The choice is a political one, but it will express itself in the legal status and regime of the oceans. It is to these legal alternatives that we will now turn.

[1] Willard Bascom, "Technology and the Ocean," *Scientific American,* September, 1969, p. 199.

[2] Marine biologists call the freely moving fish "pelagic," as distinct from two types of living organisms associated with the seabed: (1) benthonic, or sedentary organisms that live attached to the seabed; (2) demersal organisms, consisting of living organisms of the superjacent waters that are associated with the seabed because they either live on benthonic organisms or use the seabed for shelter. Among these are such economically valuable creatures as the flatfish and the shrimp.

[3] Sidney J. Holt, "The Food Resources of the Ocean," *Scientific American,* September, 1969, pp. 178ff.

[4] Charles Stewart and Giulio Pontecorvo, "Problems of Resource Exploitation: The Oil and Fishing Industries," in *Pacem in Maribus: Ocean Enterprises,* published by the Center for the Study of Democratic Institutions, 1970, p. 7. (See Bibliography, p. 126–127).

[5] Giuseppe Muscarella, "International Regime of the Seabed; Oil and Natural Gas

—Exploration and Exploitation of Petroleum in Deep Water," *Rome Symposium,* p. 109.

[6] Francis T. Christy, Jr., "Marigenous Metals, Wealth, Regimes and Factors of Decision," *Rome Symposium,* p. 115.

[7] Frank LaQue, "Deep Ocean Mining: Prospects and Anticipated Short-Term Benefits," in *Pacem in Maribus,* p. 22. An article in the *New York Times,* August 9, 1970, reported that a research vessel owned by a United States corporation had succeeded in sucking up from a depth of 915 meters, at a distance of 120 miles from the coast of South Carolina, a mixture of metallic ore nodules, air, and water. This in turn raised the possibility of bringing up nutrient-rich water from the ocean depths to higher reaches as an additional source of fish food.

[8] Christy, "Marigenous Metals," p. 129.

[9] John L. Mero, *The Mineral Resources of the Sea,* New York, 1965.

[10] Christy, "Marigenous Metals," p. 127.

[11] Holt, "The Food Resources of the Ocean," *op. cit.* (note 3), p. 178.

[12] For an excellent brief survey of these various developments, see Juraj Andrassy, *International Law and the Resources of the Sea,* New York, 1970, Chapter II.

[13] "The Living Resources of the Seabed," *Rome Symposium,* pp. 192ff.

4

The Law of The Oceans

AS a general principle of international law the freedom of the seas is less than three and one-half centuries old. It protects the surface of the seas, with the exception of territorial waters, from appropriation or exclusive use by any one state. It was logical that the doctrine should emerge in the age of exploration, when Columbus, Cabot, Vasco da Gama, Raleigh, Drake, Magellan, and others began to link the continents of America, Asia, Australia, and later, inner Africa, with Europe. The classical legal battle was fought between John Selden, who, as an Englishman in the earlier part of the seventeenth century, tried to protect the interests of a country that was not the leading naval power by advocating a closed sea, and the Dutchman, Hugo Grotius, called the father of international law, who postulated an open sea for his then-dominant maritime country. It is interesting to note that when England later gained naval supremacy, she became the leading champion of the freedom of the seas.

Except in times of war, the major maritime powers

have always been supporters of the freedom of the seas, since their navies were able to protect their commercial interests, including, in most cases, their overseas colonies, their merchant fleets, and their fishermen. Between them they were powerful enough to maintain this freedom as a general doctrine of international law. Until a few years ago they were also the principal advocates of narrow territorial waters—generally a three-mile zone of exclusive sovereignty—the distance a cannonball could be fired effectively in the days of sailing vessels. Nations with limited maritime access, like Russia, have always claimed wider territorial waters of twelve miles, and this relatively moderate extension of territorial sovereignty has now been adopted more or less universally. One after another, the traditional defenders of a narrow limit, including Canada most recently, have proclaimed a twelve-mile limit, and those countries, like the United States, that have not yet officially done so in fact come close to the same position by claiming exclusive fishery rights, police powers, and other controls over a "contiguous zone" extending to nine miles from the three-mile limit of the territorial waters. That the smaller nations should embrace the wider territorial limits doctrine is understandable, since it helps them to protect exclusive fishing rights and other privileges of sovereignty; it is somewhat less justifiable in the case of more powerful nations.

The extension of territorial water limits from three to twelve miles is a minor infringement of the freedom of the seas. However, the decisive turn away from the tenets of Grotius came with the introduction and rapid adoption throughout the world of the Continental Shelf Doctrine in the Truman Proclamation. The latter declared exclusive jurisdiction and control over "the natural resources of the subsoil and seabed of the continental

shelf beneath the high seas but contiguous to the coast
of the United States. . . ." Within a few years, one maritime state after another had expressed or made corresponding claims. Among the famous controversies of
that period was the question of Japan's traditional cultivation of oyster pearls within the continental shelf of
the north coast of Australia and the 1962 "lobster war"
between Brazil and France, which was provoked by fishing fleets from Bretagne catching *langoustes* in Brazil's
continental shelf.[1] Within thirteen years of the Truman
Proclamation it was possible to enact an international
convention,[2] which became effective in 1964, after the
twenty-second ratification, and has now been accepted
by forty-four states. However, only the parties to
the convention are bound by its terms; every other
maritime state regards the extension of its sovereignty
to the seabed and subsoil resources of the continental
shelf as a matter of customary law, with no obligation
to abide by the precise prescriptions of the Geneva Convention.

This was confirmed by the decision of the International Court of Justice on the North Sea Continental
Shelf Case (February, 1969). In 1966 a dispute had
arisen between the Federal Republic of Germany on the
one side, and The Netherlands and Denmark on the
other, concerning the boundaries of their respective
continental shelves in the North Sea. West Germany
had signed, but not ratified, the Continental Shelf Convention, in which Article VI provides that "where the
same continental shelf is adjacent to the territories of
two or more states whose coasts are opposite each other
. . . in the absence of agreement, and unless another
boundary line is justified by special circumstances, the
boundary is the median line, every point of which is

equidistant from the nearest points of the base lines from which the breadth of the territorial sea of each state is measured. . . ." In other words, the boundary line is drawn at an equal distance between coastlines.

This compelled the Court to determine the legal nature of rights to the continental shelf in the absence of an international treaty between the parties involved. The boundaries of the continental shelves of the contending parties were difficult to establish in this case, because the relevant coastline of Germany is strongly curved inward, giving her a relatively narrow continental shelf area if the equidistant principle is applied. How the Court resolved the problem is not relevant here; what matters is the court's affirmation of the existence of an *ipso jure* right of the coastal state to the continental shelf as such. It held that the "rights of the coastal State in respect to the area of continental shelf that constitutes a natural prolongation of its land territory into and under the sea exist *ipso facto* and *ab initio*, by virtue of its sovereignty over the land, and as an extension of it in an exercise of sovereign rights for the purpose of exploring the seabed and exploiting its natural resources. In short, there is here an inherent right. . . ."

The worldwide adoption of the continental-shelf concept has been so rapid and the national claims based on it so numerous that it is easy to overlook the significance of this statement. The development of a generally accepted international custom usually takes decades, or sometimes even centuries; it must be proved by practice and fortified by evidence that the practice is more than just usage and that it is the expression of a legally binding principle. However, it took less than thirteen years (the time between the Truman Proclamation and the Geneva Convention) for the extension of national sovereignty to the resources of the continental shelf to be universally accepted. As early as 1950 (five years after the Truman Proclamation) one of the most authoritative international legal scholars, the late Sir Hersch Lauterpacht, argued that the concept of the continental shelf had become part of customary international law, since consistent and uniform usage of states could be established in a short space of time:

> Any tendency to exact a prolonged period for the crystallization of custom must be in proportion to the degree and the intensity of the claims that it purports, or is asserted, to affect. With regard to the submarine areas adjacent to the coast, the assertion of sovereignty over them would constitute a drastic change in the law only if it could be shown that the international law of the sea . . . actually prohibited, instead of being merely silent on the appropriation of the seabed and the subsoil outside the territorial waters. . . . Unilateral declarations by traditionally law-abiding states, within a province which is particularly their own, when partaking of a pronounced degree of uniformity and frequency and not followed by protests of other states, may properly be regarded as providing such proof of conformity with law as is both creative of custom and constituting evidence of it.[3]

There is a bitter irony in the fact that this early affirmation of the exclusive rights of coastal states over the resources of their continental shelves was put forward by one of the most eloquent and authoritative advocates of a strengthened international legal order and the rule of law in international affairs. Lauterpacht probably underestimated the contagious effect of the new doctrine and the degree to which it would be carried out. He might well have been warned by the rapidity and eagerness with which countries adopted the doctrine of national sovereignty over air space at the end of the First World War, and the prevalence of national political and economic interests over the forces of a collaborative international order. An equally distinguished French contemporary, Professor Georges Scelle, was more prescient. In a publication of 1955,[4] he predicted that the doctrine of the continental shelf would lead to ever-increasing claims on the common domain of the high seas, both upward to embrace the superjacent waters and outward to expand the territorial sovereignties of the coastal states.

If anything, the events of the past fifteen years have exceeded Scelle's fears. We can give only the briefest survey of the developments that threaten to destroy the very foundations of the freedom of the seas. The significance of these developments goes far beyond the immediate issue of the expansion of national maritime jurisdictions; it is an ominous indication of the kind of political thinking and the strength of the economic pressures that dominate the contemporary scene.

As early as 1949 the International Law Commission, a permanent organ of the United Nations for the "promotion of the progressive development of international law and its codification," was entrusted with the task of preparing a draft treaty concerning the continental shelf. The successive drafts of the commission—which finally resulted in the Geneva Convention of 1958—illustrate the relentless expansion of a once-limited concept. The development of the treaty demonstrates how far man's technological ingenuity and scientific ability exceed his political foresight and the capacity to adapt his moral, political, and legal conduct to the technological changes that have revolutionized contemporary society.

In its first draft, in 1951, the International Law

Commission adopted a criterion that was to prove disastrous: it defined the term "continental shelf" as "the seabed and subsoil of the submarine areas contiguous to the coast, but outside the areas of territorial waters, where the depth of the subjacent waters admits of the exploitation of the natural resources of the seabed and subsoil." Influenced by many noted scholars, who proposed a definition of the continental shelf in terms of the 100-fathom, 200-meter isobath, the commission subsequently adopted a definite depth limit, stressing that the exploitation test might lead to an open-ended extension of the exclusive rights of coastal states to unlimited depths and unlimited distances from the coast. But in its final draft the commission reverted to the earlier exploitability criterion, proposed in its 1951 draft. This was adopted in Article I of the Geneva Convention, without any clear definition of the limits of national jurisdiction either in depth or in width. In vain some national delegations, such as the Dutch, proposed an absolute depth limit of 500 meters, which, although greatly exceeding the original 200-meter isobath limit, would at least have been categorical and definite. Many of the defenders of the open-ended exploitability test comforted themselves with the contention that exploitation at depths greater than 200 meters would not occur in the near future. In a paper prepared for the Geneva Conference a leading expert, the late Admiral M. W. Mouton, declared that exploitation at great depth would not be possible for at least twenty years; he believed it would take a minimum of ten years to develop structures that would permit drilling at 400 feet—a depth well within the 200-meter limit of the Truman Proclamation. Yet at the same time, there were a number of experts who foresaw the possibility of exploitation at depths of 600 meters to 1,000 meters within the next few years.

The exploitability criterion—considered at Geneva as a relatively innocuous compromise—has, in fact, proved to be the starting point for an almost unlimited outward and downward expansion of exclusive continental-shelf claims and for the virtual elimination of the concept of the continental shelf as a legally meaningful limitation of national sovereign rights.

Since the convention was subject to revision five years after it came into force, it would have been possible to adopt a definite depth limit in 1969, which could have been reviewed in the light of subsequent technological developments. This was indeed proposed by several delegations as well as by individuals, but to no avail.

In another disturbing clause the Geneva Convention allotted separate continental shelves "to the seabed and subsoil of similar submarine areas adjacent to the coasts of islands." This has had some extraordinary consequences. While the majority of the colonial possessions of the Western powers are now independent states, a number of small islands remain colonies of such countries as Britain and France. There would have been some political and geographical sense in limiting separate continental shelves to those islands that are either independent sovereign states—like Barbados, or Trinidad and Tobago in the West Indies—or to the main coast of the state of which they are a part. But since the Continental Shelf Convention failed to make any such limitation, tiny islands that are still colonial possessions—such as the French-owned Clipperton, off the coast of Mexico, and St. Pierre et Miquelon, off the Gulf of St. Lawrence, or the British possession of St. Helena in the Atlantic, where Napoleon died in exile—can claim separate continental shelves, which in some cases add enormous areas to the states to which they belong.

The convention is eloquent on the preservation of the traditional freedoms of navigation and fishing and the conservation of the living resources of the sea. It provides that the rights of the coastal state over the continental shelf do not affect the legal status of the superjacent waters and that exploitation of the continental shelf and the exploitation of its natural resources "must not result in any unjustifiable interference with navigation, fishing, or the conservation of the living resources of the sea, nor in any interference with fundamental, oceanographic, or other scientific research carried out with the intention of open publication." Coastal states are entitled only to construct and operate the necessary installations within their continental shelves in accordance with these safeguards.

However, as exploration and installations multiply, these pious reassertions of the conventional and customary freedom of navigation and fishing obviously become increasingly empty, and we are left with a question of priorities: the exclusive claims of the coastal states, or the traditional freedoms of the international community. The answer can hardly be in doubt. As we have already noted, in the Gulf of Mexico, where oil drilling has reached an intensity likely to be followed soon in other areas the relegation of shipping to "fair" shipping lanes has already forced navigation into a back seat. There is little reason to assume that any state will voluntarily forego a valuable find out of respect for the freedoms of fishing and navigation.

The Race to the Oceanbed

The consequences of the open-ended definition of the continental shelf have been many, with probably many more to come. The first, and perhaps the most obvious, response to the "exploitability" loophole has been the gradual erosion of the 200-meter isobath as the normal depth limit of the continental shelf. As oil drilling, even though it is mostly experimental at this time, and alternative methods of exploitation become available, the claims expand. Any precise limitation at a given time is apt to be short-lived, but to give some indication of the license of interpretation in this area, it may suffice to mention that at the Fifty-Third Conference of the International Law Association, held in Buenos Aires in 1968, the American branch of the International Law Association had already arrived at the suggestion of a 2,500-meter isobath, or a distance of 100 miles from the baseline of the coastal state's coastline, whichever yields the greater width. And a report by E. D. Brown, prepared for the British branch of the International Law Association, suggested a 200-mile limit, to be calculated from the base lines of the territorial sea. This last proposal suggested that "exploitability" should be understood to mean economic feasibility rather than technologically possible exploitation.

Substitution of the concept of the "continental margin" for that of the "continental shelf" represents an even more blatantly distorted interpretation of the Geneva Convention. As we saw in an earlier chapter, the continental margin comprises not only the shelf but also the continental slope and the continental rise, or in other words, the entire oceanbed up to the abyssal depth, an area representing approximately 23 per cent of the total ocean floor. Although neither the International Law Commission nor the Geneva Convention ever tried to equate the legal and geographical concepts of the continental shelf, they obviously would not have chosen this particular limitation if they had wanted to extend national jurisdiction to the continental margin. The gradual substitution of the wider for the narrower concept can only be described as a thin legalistic disguise for the unilateral assertion of national claims, in violation of an international treaty. For example, it is the term "continental margin" that is used in an interim report of the Subcommittee of the United States Senate concerned with this matter; not surprisingly, the American Petroleum Institute, which represents the interests of the American oil industry, has also claimed the continental margin as the area of exclusive national exploitation.

It was perhaps to be expected that some lawyers should have interpreted the convention as doing away with any limiting geographical criterion and giving exclusive rights to the coastal state over any part of the ocean where resources can be exploited. Thus Professor Shigeru Oda, the leading Japanese legal expert and representative to the United Nations Seabed Committee, states that "the exploitability of submarine resources at any point must always be reserved to the coastal state, which is empowered to claim the area when the depth of the superjacent waters admits of exploitation."[5] Professor F. Munch, a German authority, writes in a similar vein.[6] On this basis, all submarine areas have in fact already been theoretically divided between the coastal states at the deepest trenches in the ocean floor, and not even the continental margin provides any limit. Although Professor

Oda is in favor of a revision of the Convention, which would "remove the test of exploitability," it would obviously be far simpler to give an interpretation to the present Convention which does not in effect eliminate the concepts of the "continental shelf" and of "adjacency" in favor of open-ended national claims to any part of the seabed that is exploitable.

In his analysis of the International Court's decision in the North Sea continental shelf cases, an English scholar asserts somewhat more moderately that "the idea of a prolongation of the land domain is a legal foundation of the submerged area of national jurisdiction" and that because this concept is related to geological facts, "there would seem to be little room for doubt that the continental slope is just as much a part and a prolongation of the continental land mass as the continental shelf is." He argues that since the underlying rock structure of the shelf and the slope is identical, both are part of the prolongation of the continent over which the coastal state has exclusive jurisdiction.[7] Although Professor Jennings professes to base himself on the judgment of the International Court of Justice, he only selects those aspects that might support his advocacy of an extension of national jurisdiction beyond the continental shelf. Nowhere does the Court's judgment support indefinite prolongation of land territory into the sea. It limits national jurisdiction to the continental shelf—a concept clearly distinct from that of the continental slope and the continental rise—and stresses the concept of "adjacency," which is part of the definition of the continental shelf in Article 1 of the Convention. But Jennings even does away with the latter by stating that "adjacency comprehends the idea of appurtenance as a prolongation of the land domain."

Yet the Court had clearly said that "by no stretch of imagination can a point on the continental shelf situated, say, 100 miles or even much less, from a given coast, be regarded as 'adjacent' to it, or to any coast at all in the normal sense of adjacency. . . ." For the test of adjacency, Jennings substitutes the test of "exploitability beyond the depth of 200 meters" of Article 1 of the Convention (ignoring that this Article, too, speaks of "areas adjacent to the coast"). By this ingenious combination of different criteria, Jennings arrives at the conclusion that the continental slope and possibly the continental rise, also, are within the domain of national jurisdiction. The aforementioned "interpretations" constitute in reality de facto revisions of the Convention and are clearly designed to maximize national claims— already so greatly extended by the doctrine of the continental shelf—at the expense of international freedoms.

In a recently published book,[8] Professor J. Andrassy points out that the Geneva Convention, like the Truman Doctrine, is based solely on the concept of the continental shelf with no mention of the continental slope. Andrassy rightly asserts that the words of a legal text must be interpreted in their normal meaning, making untenable the substitution of "natural prolongation of the land mass of the continents" for the term "continental shelf." But even more important, he notes that if exclusive jurisdiction is to be allotted to the natural prolongation of the land as an inseparable part of the continent, *all* countries of the continent should have the right to a share in this extension of rights, and not only those that are situated on the coast. Since the oceanbed sediment in which there is so much interest has originated over millions of years in the land from which material is carried by rivers into the sea, it would be highly inequitable to deny a share of these resources to a country whose rivers contributed to their accumulation. This would give certain rights to landlocked countries, which are altogether excluded from the benefits that have accrued to the coastal states, as the law now stands.

It was hardly to be expected that states excluded from such an enormous extension of the wealth of certain countries would not seek some kind of compensation. The most disadvantaged of these states—the twenty-nine landlocked countries—so far have not made moves other than to support an international oceanbed regime under which at least some of the wealth of the seabed would be redistributed. A more specific response was that of those Latin American countries that, because of their steeply declining coastlines, do not enjoy an exploitable continental shelf. In the Declaration of Santiago in 1952, Chile, Ecuador, and Peru proclaimed their sole jurisdiction and sovereignty over the area adjacent to and extending 200 nautical miles from their coasts, including the sea floor and subsoil of that area. Several other Latin American states, including Argentina, subsequently made similar proclamations. This assertion of sovereignty over territorial waters to 200 miles goes further than the interest in the resources of the seabed and subsoil, with which the continental shelf and margin claims are concerned. It includes the right to exclude foreign fisheries and to control navigation of all kinds. It is hardly up to those who have claimed ever-expanding areas of the ocean for exclusive exploitation to denounce such claims—especially since the most emphatic proponents of extended national seabed control have been the major industrial and maritime nations of the world. However, in the perspective of the Grotian Doctrine of the freedom of the seas, an extension of territorial waters from the now generally accepted maximum of 12 miles to 200 miles is a disaster, turning us back three and one-half centuries to Selden's doctrine of the closed sea. The original motivation for this extension was the attempt of less developed countries, which depended greatly on their fisheries, both to compensate for the absence of exploitable continental shelves and help allay their fear of overfishing by highly mechanized foreign fishing fleets. But recently, Brazil and Uruguay, two states with extensive and highly profitable continental shelves, also proclaimed a 200-mile limit. It will not be long before almost the entire Latin American continent will have decreed a 200-mile zone of national waters around the coast.[9] Unfortunately, the erosion of the freedom of the seas does not stop here. We have already referred to the obvious danger of increasing interference with the surface of the high seas by the expanding exploitation of seabed and subsoil resources. It is almost inevitable that sovereignty over the resources of the seabed and the subsoil should extend upward, that the coastal states should regard it as their prerogative and duty to protect the installations and structures erected under their jurisdiction, and that the rights of other nations in these zones should become licenses.[10]

In April, 1970, Canada, which has played a major role in the promotion of international legal order in the post-war period, proclaimed a 100-mile zone of exclusive control in the Arctic, with the object of preventing pollution in that area. The background of this legislation is an unresolved dispute with foreign, especially American, interests in regard to the exploitation and transportation of the newly discovered oil and gas riches of Alaska and the recent passage of an American oil tanker through the Northwest Passage, which Canada claims is under its territorial jurisdiction. The substantive provisions of the new act are unexceptionable; they include extensive penalties for the deposit of waste of any type in the area under control, establish pollution-prevention officers, and impose strict liability on those responsible for any damage incurred as the result of pollution. But, like the proclamations of the Latin American states, this was a unilateral action, justified by Canada by the lack of progress in international agreements on the control of pollution and environment. The act also provides for "shipping control zones" and a parallel act authorizes the Governor-in-Council to prescribe "as fishing zones of Canada such areas of the sea adjacent to the Coast of Canada as are specified in the order. . . ." There can be little doubt that Canadian control over the 100-mile Arctic zone will soon develop into a jurisdiction barely distinguishable from that of full sovereignty. Its stated objective of pollution control and the protection of the ecology of the Arctic is entirely laudable and may prove to be a pioneer effort in wider international agreements in this vitally important area. But as it now stands it is not confined to pollution control, and constitutes a further move in the unilateral extension of national jurisdiction at the expense of the rights of the international community and the traditional freedoms of the sea.

It is a melancholy tale. Since the Truman Proclamation state practice has almost without exception moved in one direction only: the extension of national claims at the expense of international freedoms. The most important official move in the opposite direction is an announcement made by President Nixon on the United States ocean policy on May 23, 1970, and the subsequent

United States Draft Convention submitted to the United Nations, which we will discuss later.

The Status of the Deep Seabed

Except for those who in effect advocate the division of the oceans into competing national jurisdictions, the present claims stop at the end of the continental margin; they do not purport to include the abyssal depths. Of course, the latter is not likely to be a major source of mineral exploitation in the immediate future. To date, drilling experiments have not proceeded beyond 1,200 meters, and means have not yet been found of stationing crews at such depth. However, time and again the rate of technological progress has outdistanced predictions. Only a few months ago it was reported that an American ocean-research ship had succeeded in retrieving an experimental probing from a depth of 6,000 meters. The pessimistic predictions of many experts a little more than a decade ago about exploitability at depths of 600 meters to 1,000 meters have proved to be quite incorrect.

But there is another, more practical aspect to the status of the deep seabed presented by the seamounts, atolls, and shallow banks that occur outside the continental margin area. Since they lie either on or just below the surface of the water, their occupation and the erection of structures on them is immensely easier than exploration of the abyssal depths. If, as many international lawyers maintain, the freedom of the seas does not extend to the sea floor and the subsoil, such mounts or banks could be occupied, appropriated, and used to exploit adjacent areas by the occupying state, which might be a landlocked country. The oceans are free, and even if exclusive rights over the continental margin became generally accepted under international law, this would still leave three-quarters of the ocean surface open. No nation can appropriate any section of the ocean as such. But does this prohibition extend to the seabed and subsoil? This question may become crucial as deep-sea exploration proceeds and even before too long, as frustrated states may seek to occupy seamounts or atolls as a basis of operations.

The opinion has long been held that possibly portions of the seabed and definitely the subsoil under the sea are capable of effective occupation and appropriation. A leading textbook states that "there is no reason whatever for extending [the] freedom of the open sea to the sub-soil beneath its bed. On the contrary, there are prac-tical reasons—having regard to the construction of mines, tunnels, and the like—which compel, apart from the wider issue involved in the now recognized claim to the continental shelf, recognition of the fact that this subsoil can be acquired by occupation. . . ."[11] An-other writer has recently stated that while occupation of the seabed as such is prohibited, it is possible for nations to acquire a kind of possessory title, if other nations acquiesce.[12]

Most of the views affirming the appropriation of the subsoil were developed when mineral exploitation was a distant dream, but tunneling under the sea seemed more realistic. As early as 1911 Bernhard Kellermann, a Ger-man journalist, published a best-selling novel about a tunnel built under the Channel between Britain and France—a proposition revived in recent years but still far from execution. It is, however, very difficult to imagine mineral operations in the subsoil of the sea

without effective occupation of the corresponding portion of the seabed—not to speak of the necessary control over the installations below and above the sea surface, which would be connected with such operations.
It is therefore a crucial question whether the seabed is
res communis or *res nullius*. The latter theory would
permit portions of the seabed to be appropriated and
exclusive national claims to be extended even to the
open seas—particularly the sea shallows and seamounts
scattered throughout the oceans. The *res communis* theory treats the open seas as the common property of
mankind. But whichever theory is adopted remains of
purely theoretical interest unless an international authority effectively exercises control over the seabed on
behalf of the community of nations.

---

[1] In 1958, the Continental Shelf Convention stated that the "natural resources"
over which the coastal state exercises sovereignty include "organisms which, at
the harvestable stage, either are immobile on or under the seabed or are unable to
move except in constant physical contact with the seabed or the subsoil." This
includes oysters, but there was disagreement about the classification of lobsters as
seabed resources or as moving creatures of the sea. The "lobster war" was settled
in December, 1964, by an agreement that permitted a limited number of French
fishing vessels to fish for a period of five years, but obligated them to hand over
a proportion of the catch to the Brazilian group, which had sought its government's intervention.

[2] Convention on the Continental Shelf, Geneva, April 28, 1958.

[3] Hersch Lauterpacht, "Sovereignty over Submarine Areas," *British Yearbook of
International Law*, Vol. 27 (1950), pp. 393-94.

[4] *Plateau Continental et Droit International*, A. Pedone, 1955.

[5] Shigeru Oda, "Proposals for Revising the Convention on the Continental
Shelf," *Columbia Journal of Transnational Law*, Vol. 7 (1968), p. 9.

[6] *Archiv des Volkerrechts 180* (1959/1960).

[7] R. Y. Jennings, "The Limits of Continental Shelf Jurisdiction," *International
and Comparative Law Quarterly* (1969), p. 819.

[8] *International Law and the Resources of the Sea* (1970).

[9] In August, 1970, at a congress held in Peru, fourteen Latin American countries declared that all nations have the right to claim as much of the sea and the
seabed near their coasts as they deem necessary to protect their actual and potential offshore wealth. Two landlocked countries, Bolivia and Paraguay, voted

against the declaration, and so did oil-rich Venezuela. Cuba and Haiti did not attend. The major Caribbean states, Jamaica, Trinidad and Tobago, and Barbados abstained or left the conference, having expressed concern over the consequences of the declaration for countries that are close to each other. (*New York Times,* August 16, 1970.)

[10] In 1967, Dr. John Craven described the "trend of law" as "the careful extension of municipal jurisdiction seaward along the seabed and from thence to the vertical column above and the subterrain below." Paper presented at First Conference on Law, Organization and Security in the Use of the Ocean (Ohio State University, 1967), p. C35.

[11] Ludwig Oppenheim, *International Law* (Lauterpacht, ed., London, 1955), Vol. 1, p. 630.

[12] Ian Brownlie, *Principles of International Law* (1966), p. 206.

5

Military and Strategic Uses of The Seabed

WE have seen how the steadily expanding claims of coastal states to ever-increasing portions of the oceanbed inevitably entail corresponding efforts to protect the physical and financial investments made by their nationals. It would be naive to think that the exclusive exploitation and other uses of certain portions of the seabed would not affect the rights to the surface waters above them and to any structures that reach from the surface down to the ocean floor. It would be even more naïve to believe that the exclusive controls claimed by states—be it in the form of 200-mile territorial water zones, exclusive exploitation rights to the continental margin, or exclusive environmental controls —would not extend to military considerations. A state that claims sovereignty over the continental shelf or continental margin, or erects structures on a sea bank, is not likely to tolerate the military proximity of a potentially hostile state, let alone the use of such structures by them.

The seas have long been the theater for naval battles as well as for peaceful commerce. Freedom of the seas has included the freedom of all nations to fight as well as to trade with each other. This has proved to be a great advantage to the major maritime powers in time of war, enabling them to protect their vital supply lines, to intercept supplies needed by their enemies, and to exercise pressure on neutral states by reducing their commerce with the enemy or cutting it off altogether. The naval superiority of Britain and the United States played a vital role in the ultimate victory of the Allies over the relatively landlocked Germany and her continental allies in both World Wars. Although future wars may see the decisive impact of naval power displaced by air forces and missiles, this might not apply to the limited warfare that characterizes, for example, the present hostilities in the Middle East, where the naval strength of the United States and the Soviet Union are vital factors in the balance of forces. The growing accessibility of the oceanbed now adds a new dimension to the military balance of power and may prove even more decisive than the battle for air supremacy.

There are two broadly different conceptions of the military use of the seabed. The first gives the oceanbed the same legal status as the superjacent waters—as in the case of internal waters and the territorial sea. This applies the same freedom to the seabed and the subsoil as to navigation on the high seas. Since the high seas have always been used in times of war as in peace by the warships of all nations, and in recent years even for nuclear test explosions—notably by the United States and France—it would follow that countries are equally free to use the oceanbed for naval purposes.

A second and opposite view reasons that since the

ocean depths were until quite recently inaccessible to man, no usages or customs have developed with regard to the oceanbed, which must be considered a legal vacuum like outer space, and for which laws have to be developed by practice and international treaties. This attitude was adopted by the Chilean delegate at the twenty-third session of the United Nations General Assembly. He maintained that outside the limits of national jurisdictions there were no rules of international law governing the uses of the seabed, and the legal regime remained to be developed by international legislation.

In this confused legal situation the United States and the Soviet Union have led some intense maneuvering over the uses of the seabed for military purposes. Ever since the submarine became a major weapon in World War I and World War II, the use of the oceans for military purposes under the surface and down to a very considerable depth has been taken for granted. The strategy of both superpowers is to a large extent based on nuclear-propelled submarines equipped with long-range missiles. The invisibility and mobility of such deadly arsenals—now to be increasingly equipped with multihead missiles—makes them virtually invulnerable, while they pose a lurking threat of near-total destruction to the enemy. The tremendous increase of interest shown by United States authorities in the military aspects of the ocean floor in recent years is indicated by the 1969 appropriation of $516 million for oceanographic programs as opposed to its allotment of only a few million dollars ten years earlier.[1]

In the prolonged discussions of the eighteen-nation United Nations Disarmament Committee, neither the United States nor the Soviet Union has proposed any limitation of the use of the seas for submarines. As long as the United States had one-sided superiority in this weapon, the Soviet Union might well have pressed such a proposal, but in recent years, the U.S.S.R. has itself built up a formidable force of nuclear-powered and missile-equipped submarines, which constitutes a potential menace to the United States in the Atlantic, the Pacific, and the Mediterranean. Both states have, however, suggested certain limitations to the military uses of the seabed. The Soviet Union has presented a draft treaty that would prohibit the use of the seabed and the subsoil of the ocean floor for any military purposes. The prohibited area would encompass the seabed and subsoil beyond a twelve-mile zone measured from the base lines used to define the limits of territorial waters. The United States has advocated a considerably more limited international agreement, which would prohibit only the emplacement or fixing of nuclear weapons or other weapons of mass destruction on the seabed. The apparent reason for this is that the United States and its allies attach great importance to the installation of "listening" systems on the ocean bottom to be used for anti-submarine warfare. The U.S.S.R. seems to be less dependent on such systems.[2] The two superpowers finally agreed on a joint minimum proposal that would prohibit the installation of nuclear and other mass destruction weapons on the oceanbed. This was approved by twenty-four members—all except Mexico—at the Geneva Disarmament Talks on September 3, 1970. On November 17, 1970, the Political Committee of the United Nations approved the draft treaty by a vote of 91 to 2. However, the prohibition of fixed oceanbed installations would do little to diminish the danger of nuclear war and destruction from the sea, as long as mobile underwater warfare and weapons are allowed.

But it is not only missile-carrying and nuclear-propelled submarines that constitute a deadly danger as the principal potential weapon of mass destruction in another world war. What is no less dangerous—just as it is highly promising for the development of peaceful commerce—is the increasing accessibility of the oceanbed for transportation. Dr. John P. Craven,[3] one of the world's leading authorities on the subject and until recently chief scientist in the Strategic Systems Project Office of the United States Navy, observes that the "sea system" (the use of the surface of the oceans) has always been subject to certain limitations and handicaps, of which he lists six: (1) the perils posed by the changing conditions of the wind and the sea; (2) the impossibility "to make a landfall at an arbitrary portion of the coast for transfer of personnel or cargo" under moderate or modest sea conditions; (3) the limitations of speed on the seas; (4) the exposure to optical and electromagnetic spectra; (5) accommodation of large volumes and tonnages, limited by draft and harbor conditions; (6) accessibility of seaborne vessels and installations to aircraft or airborne vehicles.

Some of these factors present advantages as well as perils for activities such as the movement of large ships and cargoes across the seas. But in times of war the perils outweigh the benefits. The ocean floor therefore presents certain unique and undeniable military advantages over the surface and superjacent waters. First, operations on the seabed would be independent of the conditions of wind and sea. Second, it would be possible under most conditions to transfer personnel and cargo anywhere along any coast. Third, any transit system on the oceanbed is essentially free of wave drag. Fourth, large volumes and tonnages could be accommodated, although in this repect there are structural limitations. Above all, oceanbed systems are practically inaccessible to optical and electromagnetic spectra, as well as to aircraft, and virtually invulnerable to attack from the air.

Another expert, who favors the controversial antiballistic missile defense,[4] has suggested that listening posts on the ocean floor may not be as vital to the United States as they are now held to be. He foresees the rapid development of air-cushioned surface ships with speeds ranging to 100 knots for medium-sized ships of up to a few thousand tons by the end of the century. Since there is no prospect of a corresponding acceleration of submarine speeds, surface ships would be much safer in the future.

Giant transport planes (such as the trouble-plagued C–5A, which recently made its debut in Vietnam) have begun to provide another alternative for the movement of troops across the oceans. Such giant aircraft could in due course give decisive logistic support in major wars, even at remote distances, and thus reduce the importance of submarine warfare.

Such developments are predicted for the year 2000. But what of the critical thirty intervening years? Given the present combination of technological development and political tensions, three decades are a long time. For the moment it seems almost certain that none of the major powers would forgo the advantages of using the seabed, beyond the banning of nuclear installations. Even if the United States and the Soviet Union were to develop sufficient trust in each other's promises to mutually abandon the advantages of submarine and seabed mobility, China, with her growing technological and nuclear capability, remains tragically outside the main web of international relations and commitments.

France has refused to accede to the treaties banning surface nuclear tests and the proliferation of nuclear weapons; and most of the other middle-sized naval powers are no more likely to forego immediate strategic advantages for the sake of long-range objectives. The political factors that have paralyzed any meaningful disarmament over the past few decades are just as likely to prevail in the new dimension of oceanbed warfare. Once again, the strategic need of mutually deterrent power will be used as justification for weapons build-up. In the meantime, it is likely that coastal states will gradually extend their control over the continental shelves to include any and every kind of seabed installation and operation. While the Continental Shelf Convention forbids "any unjustifiable interference with navigation, fishing, or the conservation of the living resources of the sea . . . ," it does permit the costal state "to construct and maintain or operate on the continental shelf installations and other devices necessary for its exploration and the exploitation of its natural resources." The coastal state is also permitted to establish safety zones up to a distance of 500 meters around its installations and take any measures necessary for their protection. No state powerful enough to protect its coasts would permit another state to construct a potentially strategic transportation or other system on the floor of its continental shelf. And it is probably only the limitations of naval and air power that have so far prevented coastal states from proclaiming a ban on submarine movements within their continental shelves.[5]

In sum, the prospects for an effective international

demilitarization of the oceanbed are only slightly less gloomy than those for the abolition of submarine warfare, or indeed of any naval warfare. Disarmament is essentially a matter of faith and therefore a function of the political attitudes and relations between nations. A third world war has so far been avoided not because of disarmament but because of the present "balance of fear"—the certainty of the major military powers that another world war would lead to total destruction. And while there have been tough and militant leaders, since Hitler there has not been a madman at the helm of a major state willing to risk the total destruction of his own country for the sake of the pursuit of power. This, however, is a precarious foundation for a lasting peace.

Equally little comfort is to be derived from the treaty drawn up in January, 1967, under the leadership of the United States, the Soviet Union, and Great Britain to lay down certain principles "governing the activities of states in the exploration and use of outer space." The treaty states that "outer space, including the moon and other celestial bodies, is not subject to national appropriation by claims of sovereignty. . . ." It also obligates the parties "not to place in orbit around the earth any objects carrying nuclear weapons or any other kinds of weapons of mass destruction, install such weapons on celestial bodies, or station such weapons in outer space in any other manner." It forbids the establishment of military bases and other military installations on celestial bodies.

Even if this treaty continues to be observed when the two leading space powers have put permanently manned space stations in orbit, the use of the oceanbed offers more immediate and familiar short-term military advantages. There is perhaps still a slim chance of preventing the use of the oceanbed for military purposes, so long as no country has actually begun to construct installations or otherwise include its use in its strategic planning. Once such steps have been taken it will be infinitely more difficult to reverse directions.

Scientific Research, International Co-operation, and the Control of the Oceanbed

There has never been a greater need for expanded and intensive scientific co-operation between nations on the many interrelated aspects of the oceans and their resources than now. "In the field of non-living resources, there have to be geological and geophysical surveys of the continental margin, worldwide surveys of the topography of the sea floor, deep drilling at selected sites, both within and outside the continental margin, examination of the magnetism of sub-oceanic rocks, studies of the processes operating near the crests of mid-ocean ridges, studies of the relationship between land and sea structures in the trench arc system that surrounds the Pacific Ocean,[6] and surveys of the distribution and composition of manganese nodules in deep-sea areas."[7]

With regard to living resources, their concentration will have to be determined by systematic surveys in productive regions (with the help of acoustic and other techniques). A study of the entire ecosystem of fishing is needed as a basis for any rational regulation of the rate of fishing. There has already been much discussion of the intensification and rationalization of fishing through what some scientists call "fishing ranches" or "aquaculture" areas.

All these surveys are closely connected with additional intensive study of ocean circulation and other still only partly explored aspects of the way in which water currents, winds, temperature changes, submarine rock layers, and other aspects of oceanography affect the living resources of the sea. In an era of explosively expanding populations this matter is of more vital importance to mankind than ever before.

But this, like so many other aspects of the freedom of the seas, is gravely threatened by the expanding claims of the different coastal states.

The Geneva Convention on the Continental Shelf deals with this matter in Article 5 (1) (8):

Art. 5. (1) The exploration of the continental shelf and the exploitation of its natural resources must not result in any unjustifiable interference with navigation, fishing or the conservation of the living resources of the sea, nor result in any interference with fundamental oceanographic or other scientific research carried out with the intention of open publication.

(8) The consent of the coastal State shall be obtained in respect of any research concerning the continental shelf and undertaken there. Nevertheless, the coastal State shall not normally withhold its consent if the request is submitted by a qualified institution with a view to purely scientific research into the physical or biological characteristics of the continental shelf, subject to the proviso that the coastal State shall have the right, if it so desires, to participate or to be represented in the research, and that in any event the results shall be published.

Both sections leave a good deal of discretion to the coastal state for the areas over which it claims exclusive or predominant jurisdiction, and it is highly unlikely that any state will give permission to carry out fisheries research to other states within their zones of exclusivity, whatever they might be.

Another problem is presented by the difficulty of distinguishing between "purely scientific research into the physical or biological characteristics of the continental shelf" and less "purely scientific" purposes—notably examination of the topography for military purposes. Unfortunately, as in the field of nuclear energy, the same kind of scientific research may have peaceful or military purposes and applications. Moreover, since the installation of various structures tends to give rise to claims of exclusive control and protection, the area open to free scientific research is likely to be further reduced. All this, of course, applies only to areas of "national jurisdiction." But, as we have noted throughout this study, the limits of national jurisdiction have been uncertain ever since the absolute depth limit of 200 meters for the continental shelf was diluted by the vague test of exploitability and major coastal states proceeded to give exploratory and exploitation licenses well beyond that depth, within the continental rise or the continental slope.

As long as the limits of exclusive national jurisdiction remain as ill-defined as they are at the present time, and as long as the decision whether to grant or withhold consent rests with the coastal state, we have to expect that in the prevailing climate of military confrontation and political distrust, national suspicions and rivalries will prevail over the needs for international co-operation. Any attempt to restrict oceanographic or biological research in the area of the continental shelf is all the more deplorable, since the bulk of exploitable fish and their breeding grounds are in that area.

This state of affairs, compounded with all that has been discussed earlier, points up the urgency of redefining the areas of national control and creating some kind of international authority that would have general jurisdiction over the economic, military, and scientific uses of the oceanbed beyond the new limits of national sovereignty.

[1] Dr. Jens Evensen, "The Military Uses of the Deep Ocean Floor and Its Subsoil —Present and Future," *Rome Symposium*, p. 539.

[2] This view is stated by Donald G. Brennan, an expert on national security affairs in the Hudson Institute, one of the leading strategic analysis centers of the United States (*Rome Symposium*, p. 513).

[3] John P. Craven, *"Res Nullius de facto.* The Limits of Technology," *Rome Symposium*, pp. 520ff.

[4] Donald G. Brennan, "The Case for Missile Defense," *Foreign Affairs,* April, 1969, and *Rome Symposium*, p. 573.

[5] In the case of the United States and the U.S.S.R., this is counterbalanced by their mutual interest in having their submarines operate as close as possible to the limits of the territorial waters of the other side.

[6] The trenches occupy 1 per cent of the total sea area; they are located on the margins of the oceans and are accompanied by island arcs on their inner sides (for example, the Aleutian, Kuril, Japan, etc.).

[7] Roger Revelle, "Scientific Research on the Seabed," *Rome Symposium,* p. 650.

6

Toward an International Oceanbed Control

THE countries involved in the race for the progressive partition of the oceanbed have unequal handicaps. Only a small number of states is favored geographically and technologically; these states tend to attract a number of client states who hope to profit from their association with the major corporate enterprises of the advantaged countries.

Every year more exploratory licenses are granted by the major maritime and technologically advanced powers. Obviously, every state with the necessary technological equipment and scientific know-how will seek to extend the area over which it controls the granting of exploratory permits, if only to counter corresponding moves by other coastal states. Already on the American continental slope exploration permits have been granted for areas extending as far as 300 miles from the coast at depths ranging from 198 meters to 1,525 meters, and the United States Draft Convention of August, 1970, specifically protects investments made up to the date when the treaty will come into force. Thus for each year that elapses without the establishment of some international control the domain of vested interests expands, and the area to which an international regime might be applicable shrinks.

Naturally, the individuals and organizations who oppose any effective international control and seek to maximize the area of exclusive national jurisdiction describe as "premature" the establishment of such an authority. In 1967 a subcommittee of the United States House of Representatives reported that it would "be precipitate, unwise, and possibly injurious to the objectives that both the United States and the United Nations have in common, to reach a decision at this time regarding a matter that vitally affects the welfare of future generations." The committee also specifically opposed United States support for "the vesting of title to the seabed, the ocean floor, or ocean resources, in any existing or new organization." It concluded that "hasty action in this field could create more problems than it will solve or avert." In an interim report issued in July, 1968, the Committee on Deep Sea Mineral Resources of the American branch of the International Law Association expressed its opposition to any "supranational authority," with the power to grant or deny concessions. In order for its full implications to be appreciated, this position has to be read in conjunction with the report issued by the National Petroleum Council in 1969, which stated its "firm and carefully considered conclusion . . . that the United States, in common with other coastal nations, now has exclusive jurisdiction over the natural resources of the submerged continental mass seaward to where the submerged portion of that mass meets the abyssal ocean floor. . . ."

The nationalistic approach that has been predominant in the attitudes of legislators, the major industrial interests, and a considerable number of legal scholars clearly seeks to minimize the area of international jurisdiction, or to exclude altogether any international control over the oceanbed. Their minimum demands extend the exclusive jurisdiction of the coastal states to the edge of the continental margin, so that any possible international commission would be confined to the abyssal depths. While a permanent solution is delayed, exploratory and exploitation licenses are issued to that limit in the expectation that licenses or titles once granted would not be abandoned by the state concerned.

Although the problem of the future exploitation of deep-sea resources and the need for some international authority have been stated in previous decades, it was Ambassador Arvid Pardo's speech to the United Nations in 1967 that started a worldwide discussion of the subject. After an extensive survey of the growing importance of seabed resources and of the impact, manner, and control of their exploitation on man's political and economic future, Mr. Pardo introduced a draft resolution whose most important aspects were: (1) the exclusion of the seabed and the ocean floor "beyond the limits of present national jurisdiction" from national appropriation; (2) the establishment of an international agency to regulate, supervise, and control all oceanbed activities beyond the limits of national jurisdiction.

The first proposition was spelled out in the following four principles:

1. The seabed and the ocean floor underlying the seas

beyond the limits of present national jurisdiction are not subject to national appropriation in any manner whatsoever.

2. The exploration of the seabed and the ocean floor underlying the seas beyond the limits of present national jurisdiction shall be undertaken in a manner consistent with the principles and purposes of the Charter of the United Nations.

3. The use of the seabed and the ocean floor underlying the seas beyond the limits of present national jurisdiction and their economic exploitation shall be undertaken with the aim of safeguarding the interests of mankind. The net financial benefits derived from the use and exploitation of the seabed and the ocean floor shall be used primarily to promote the development of poor countries.

4. The seabed and the ocean floor underlying the seas beyond the limits of present national jurisdiction shall be reserved exclusively for peaceful purposes in perpetuity.

The functions of the proposed international agency were describd as follows:

1. To assume jurisdiction as a trustee for all countries over the seabed and the ocean floor underlying the seas beyond the limits of present national jurisdiction.

2. To regulate, supervise, and control all activities thereon.

3. To ensure that the activities undertaken conform to the principles of the provisions of the proposed treaty.

The Pardo proposals triggered a series of conferences, symposia, monographs, and further proposals, both official and unofficial. Before discussing the main aspects of the Pardo Resolution, we will briefly analyze its two most important official sequels: (1) the steps taken by the United Nations itself; (2) the announcement made by President Nixon on May 23, 1970, regarding United States ocean policy, followed by a draft convention presented to the United Nations Seabed Committee in August.

In December, 1967, the General Assembly of the United Nations established an *ad hoc* committee consisting of thirty-five members "to study the scope and various aspects of the Pardo idea." The special committee met three times and submitted a report that indicated unanimous approval of the President's call for an international decade of ocean exploration and general agreement on the establishment of a submarine area beyond the limits of national jurisdiction. But there was sharp division over the question of creating an international agency to control oceanbed activities. With the apparently accidental exception of Japan, all the major industrial powers, both capitalist and socialist, either voted against the setting up of a permanent committee to study the establishment of an international oceanbed authority, or abstained from voting, which amounts to the same thing, though with less candor. This vote clearly showed that none of the major powers was prepared to commit itself to an international control authority.

In December, 1968, as a result of the special-committee report, the General Assembly established a permanent committee composed of forty-two states for the peaceful uses of the seabed and the ocean floor beyond the limits of national jurisdiction; 112 nations voted favorably, none against, and there were seven abstentions, which included the Soviet Union and the two Soviet Republics with separate United Nations votes. The

committee was instructed: (1) to study the legal principles and norms that would promote international co-
operation in the exploration and use of the seabed and
the subsoil beyond the limits of national jurisdiction;
(2) to study the means of encouraging the exploitation
and use of the resources of this area in the light of foreseeable technological development and economic implications, "bearing in mind the fact that such exploitation should benefit mankind as a whole"; (3) to review
and stimulate the exchange and widest possible dissemination of scientific knowledge on the subject; (4) to
examine proposals to prevent marine pollution that may
result from resource exploration and exploitation.

In a further resolution the committee was instructed
to study the question of reserving the seabed for exclusively peaceful purposes. Finally, the General Assembly
requested the Secretary-General to undertake a study of
the question of appropriate international machinery,
"taking into special consideration the interests and
needs of the developing countries." The Seabed Committee, which divided itself into subcommittees to deal
with economic, technical, and legal problems, submitted
a report to the first committee of the General Assembly
in 1969, which resulted in certain resolutions in the
Twenty-fourth Session. Since neither the Seabed Committee nor the General Assembly have as yet arrived at
any specific operational recommendations, it may suffice
to briefly summarize the general tenor of the recommendations.

The legal subcommittee of the Seabed Committee
agreed that some area of the seabed should remain outside the limits of national jurisdiction and appropriation, but there was no concurrence on the major problem of establishing a precise boundary for this area. The

counterpart concept that the seabed and subsoil beyond the limits of national jurisdiction "are the common heritage of mankind" was "widely supported but not acceptable to all." There was also general agreement on the reservation of the seabed for peaceful purposes but not on the geographic limits for the application of this principle, or on "the scope of the prohibition of activities." Again, some general consensus emerged "on the need for the establishment of a regime . . . on the use of the resources for the benefit of mankind irrespective of the geographical location of states and taking into account the special interests and needs of the developing countries." But the character, scope, and area of such a regime were not defined. The principles of freedom, nondiscrimination, international co-operation, and noninterference with fundamental scientific research were affirmed, but a distinction was made between "scientific research" and "commercial exploration."

The General Assembly affirmed these general and in many ways ambiguous expressions of opinion. It also requested the Secretary-General "to ascertain the views of Member States on the desirability of convening at an early date a conference on the law of the sea to review the regimes of the high seas, the continental shelf, the territorial sea and contiguous zones, fishing and conservation of the living resources of the high seas, particularly in order to arrive at a clear, precise and internationally accepted definition of the area of the sea-bed and ocean floor which lies beyond the limits of national jurisdiction, the type of international regime to be established for that area."

In another resolution adopted at the same meeting (in January, 1970), a divided Assembly declared that

. . . pending the establishment of the aforementioned international regime:

(a) States and persons, physical and juridical, are bound to refrain from all activities of exploitation of the resources of the area of the sea-bed and ocean floor, and the subsoil thereof, beyond the limits of national jurisdiction;

(b) No claim to any part of that area or its resources shall be recognized.

The major powers dissented from this resolution—whose significance is in any event limited as long as "the limits of national jurisdiction" are not clearly defined. But the refusal of the major powers to assent to any moratorium indicates clearly enough their reluctance to limit the exploration and exploitation of areas they consider to be within the limits of national jurisdiction, pending the conclusion of any international treaty that would define and limit such jurisdiction.

The General Assembly invited the Seabed Committee to continue its work and to submit a draft declaration on the principles of international co-operation in the exploration and use of the seabed and subsoil, and on the economic and technical conditions that are to govern the exploitation of their resources. Held in the summer of 1970, the Geneva session of the Seabed Committee ended in complete deadlock. The Latin American states were unwilling to abandon or restrict their claims to absolute sovereignty over a 200-mile zone. The U.S.S.R. displayed its customary aversion to an international authority equipped with effective powers. The United States draft treaty did not even reach the stage of serious discussion.

Any appraisal of the briefly outlined work of the United Nations in this area must take the following points into consideration: (1) it has not yet reached the stage of any concrete operative proposals; (2) any resolution that might be passed by the General Assembly would not be directly legally binding upon the member states but would have only moral force. The resolutions might at best state certain principles, from which it would henceforth be somewhat more difficult for any one member of the United Nations to depart than before; (c) any legal commitment would have to follow from a treaty, or a series of treaties, as they might result from future sea-law conferences. But this would be a prolonged and highly complicated process, and it is doubtful that it would attain even the minimum objective of revising the First Article of the Geneva Convention in order to fix the boundaries of the continental shelf. In the meantime, it is to be feared that within the limits of technological and commercial feasibility, the coastal states will proceed with the utmost expansion of national claims and interests.

The Nixon Announcement on United States Oceans Policy

Since the effective decision-making power still rests overwhelmingly with each country, and in this area particularly with the major industrially and technologically advanced coastal states, President Nixon's announcement of the United States oceans policy and the subsequent draft convention are of special importance. But in assessing their implications, we must bear in mind that the President of the United States, however powerful, cannot make a law without the approval of Congress, and that until now the preliminary studies and statements formulated both by the Senate and the House of Representatives have been overwhelmingly in favor of the expansion of United States claims and against any effective international authority limiting or controlling such expansion. In order to become effective between nations the draft convention would first have to pass the hurdles of congressional legislation and then obtain sufficiently widespread acceptance to lay the foundation for an international treaty. Such a treaty would have to be approved by a two-thirds majority of the Senate to become binding on the United States, and at present it cannot be assumed that the policy statement represents Government policy in spite of the weight lent to it by the President's name. The sharp differences of opinion between the Department of State, the Department of Defense, and the Department of the Interior have already been noted, and in his summary accompanying the Draft United States Convention, the legal adviser to the Department of State admitted that it does not necessarily represent the definitive views of the United States Government. The United States initiative nevertheless constitutes the first modest attempt to redirect a race that during the past quarter of a century has been entirely in one direction: the outward and downward expansion of national claims to the seas at the expense of international freedoms.

First, the President of the United States affirms that decisions of "momentous importance" face mankind about "whether the oceans will be used rationally and equitably and for the benefit of mankind or whether they will become an arena of unrestrained exploitation and conflicting jurisdictional claims in which even the most advantaged states will be losers." But such sentiments have been uttered before (by President Johnson in 1966), and more important are the concrete proposals to stem the "unrestrained exploitation and conflicting jurisdictional claims" spelled out in the draft convention of August 3, 1970.

The first welcome proposal is to confine the continental shelf proper to a depth of 200 meters, which exceeds only slightly the 100-fathom (183-meter) limitation of the Truman Proclamation, to obligate all nations to renounce national claims beyond this depth, and to agree to regard the resources of the high seas beyond as "the common heritage of mankind."

Second, the United States proposes an international regime for the exploitation of seabed resources beyond the limits of national jurisdiction. An International Seabed Resources Authority is to have power to collect mineral royalties for international community purposes, particularly economic assistance to developing countries. It is also to establish rules to protect the other uses of the ocean and to prevent pollution. On the other hand, it is "to assure the integrity of the investment necessary for such exploitation."

A return to a strict 200-meter depth limit of the continental shelf and the internationalization of the oceanbed and subsoil beyond would be nothing less than revolutionary. It would return national claims to the status of twenty to twenty-five years ago and greatly enlarge not only the mineral and other seabed resources available to mankind, but also the institutional and operational authority of the international community. The United States proposal does not go that far, and it would presumably have little hope of adoption in Congress, or by the other major coastal powers, if it did. It therefore compromises between the clear-cut "nationalist" and "internationalist" approaches by proposing that coastal nations act as trustees for the international community in an international trusteeship zone consisting of the continental margins beyond the depth of 200 meters off their coasts. For this "intermediate zone"—a concept introduced by Professor Louis Henkin of Columbia University in 1968—the Nixon statement and the draft convention go along with the "continental margin" concept sponsored by the American Petroleum Institute, various congressional committees, and both American and non-American legal scholars. The crucial question, therefore, is what is meant by "international trusteeship."

The draft convention submitted by the United States to the United Nations Seabed Committee in August, 1970, bears out and clarifies the general statement made by President Nixon. In the international trusteeship area, Article 27 says that each coastal state shall be responsible for:

a. Issuing, suspending and revoking mineral exploration and exploitation licenses;

b. Establishing work requirements, provided that such requirements shall not be less than those specified in Appendix A;

c. Ensuring that its licensees comply with this Convention, and, if it deems it necessary, applying standards to its licensees higher than or in addition to those required under this Convention, provided such standards are promptly communicated to the International Seabed Resources Authority;

d. Supervising its licensees and their activities;

e. Exercising civil and criminal jurisdiction over its licensees, and persons acting on their behalf, while engaged in exploration or exploitation;

f. Filing reports with the International Seabed Resource Authority;
g. Collecting and transferring to the International Seabed Resource Authority all payments required by this Convention;
h. Determining the allowable catch of the living resources of the seabed and prescribing other conservation measures regarding them;
i. Enacting such laws and regulations as are necessary to perform the above functions.[1]

Appendix A—to which Article 27 refers—distinguishes between nonexclusive exploration licenses and exclusive exploitation licenses. The latter alone give the right to undertake deep drilling for exploration or exploitation. As the authorizing party, the coastal state must certify the operator's financial and technical competence and require him to conform to the terms of the license. There are detailed provisions concerning matters such as the size of the blocks to be licensed, the scale of fees to be charged for exploration and exploitation respectively, the submission of work plans and data under exploitation licenses and production plans prior to beginning commercial production. Particularly important from the standpoint of the interests of the international community is the proportion of between 50 per cent and 66-2/3 per cent of the revenues derived by the trustee state from license fees, rentals, and other proceeds that would be handed over to the International Seabed Resource Authority, to use "for the benefit of all mankind," and particularly to promote the economic advancement of developing states' parties to the convention, irrespective of their geographic location. Article 5 of the Draft further provides that:

A portion of these revenues shall be used, through or in cooperation with other international or regional organizations, to promote efficient, safe and economic exploitation of mineral resources of the seabed; to promote research on means to protect the marine environment; to advance other international efforts designed to promote safe and efficient use of the marine environment; to promote development of knowledge of the International Seabed Area; and to provide technical assistance to Contracting Parties or their nationals for these purposes, without discrimination.[2]

A second important feature of the draft convention as spelled out in Appendix A, provides for the liability of both the operator and the authorizing state for damages to other uses of the marine environment and for the cost of restoration. Moreover, in accordance with the standards laid down in the Convention for Seabed Exploration, the licensing state will have the power to revoke a license in the case of violation of any of the conditions imposed upon the licensee.

The United States draft generally clarifies the points left vague in the Nixon statement. "International trusteeship" is not just a thin disguise for unlimited power of the coastal state to do what it likes in the intermediate zone apart from an arbitrary financial handout to the international community. It subjects the intermediate zone and the coastal state responsible for its administration to the general standards of the convention and provides for a minimum 50 per cent share for the international community in the revenues derived from this zone. Anything less would, of course, leave the area between the continental shelf and the edge of the continental margin open to the most divergent practices, greatly increase the danger of pollution and destruction of marine life, and threaten the other international uses of the sea.

From the point of view of international community interest, a major drawback of the United States draft is its consistency with the attitude taken in United Nations debates by the developed countries in refusing to impose a moratorium, and its specific protection of investments made until the proposed treaty comes into force. Since such a treaty is at best some years away, it means that with every day that passes the scope and area of licenses granted by costal states expands, and the area left to international control—if and when a treaty comes into force—is proportionately reduced.

British and French Government proposals submitted to the United Nations Seabed Committee in the summer of 1970 generally support the United States Government's attitude to the major aspects of seabed control. They are, however, far less detailed and definite than the United States Draft Convention. In particular, they fail to set a strict limit for the continental shelf—without which it is impossible to define the limits of national jurisdiction. Instead, the British working paper simply states that "the agreement should define the area in which the [international] regime is to apply," and that the international body to be created should administer "appropriate parts of the regime. . . ." The British and French proposals generally agree on the need for an international agreement to lay down strict standards with regard to pollution control, conservation of natural resources, prevention of unjustifiable interference with navigation and fishing, and the allocation of some portion of the revenues derived from licensing fees to the landlocked and developing countries. Unlike the United States draft, both the British and French proposals would have licenses issued to states only, which in turn would sub-license operators within the area and be responsible for compliance with minimum standards of efficiency, such as conservation and pollution prevention. The French proposal makes a distinction between two types of oceanbed operations: the first—mining with mobile equipment—would be open to any state on a nonexclusive basis and subject simply to registration with an international organization and compliance with international regulations safeguarding the freedom of the seas, protection against pollution, and so on; the second—operations that entail the use of fixed installations —would be far more closely controlled by the international authority, which would grant licenses to states for specific areas and a given period of time. The French proposal also lays down that "no state may claim a monopoly of the areas adjacent to its continental shelf." The French proposal would not put the assessment and collection of part of the revenues for the benefit of development and landlocked countries in the hands of the international authority, but would make states levy a tax on sub-licensed companies for this purpose.

Thus, at least the British and French Governments give general support to a meaningful international regime. But without early agreement on the precise limits of national jurisdiction—as they are proposed only by the United States Draft Convention—the jurisdiction of any international oceanbed authority will remain shadowy and contract from year to year, while states continue to expand the areas of their national jurisdiction.

The Limits of National Jurisdiction

The Maltese proposals, the United Nations debates, and the United States proposals all agree on the necessity of some kind of international control authority over an area of the oceanbed determined to be "beyond the limits of national jurisdiction." The first task then is a redefinition of the limits of national jurisdiction that will draw precise boundaries between the areas of national and international control. At the very least, this implies total rejection of the division of the oceanbed among coastal states according to the equidistance principle. An international control area might be acceptable to those who advocate exclusive national jurisdiction up to the edge of the continental margin. But, as we have already seen, the exploitation of deep-sea resources, generally at depths exceeding 5,000 meters, will be of relatively little importance in the future and would make a negligible contribution to any adjustment of revenue between developed and developing countries. Dr. Frank LaQue has estimated that the revenue from taxes on the profits derived from the exploitation of deep-sea ocean metals would represent only a little more than .025 per cent of the world gross national product and only about .2 per cent of the gross national product of the developing nations (based on 1967 figures). If divided equally among the 1.6 billion people of the developing countries, it would amount to forty-one cents a head.[3] It is possible that others might make somewhat more optimistic estimates, but there can be no serious disagreement that an international ocean regime limited to the abyssal depths of the oceanbed would hardly be worth setting up. Its resources would be so negligible, compared with those derived by the major coastal states from the continental margins, that it would be likely to increase rather than reduce international tensions and the sense of injustice felt by the underprivileged nations.

The alternative offered by the United States proposal

does reintroduce a strict depth limit for the continental shelf. Its potential value as a basis for an internationl oceanbed regime, which would have limited authority over the area between the continental shelf and the remainder of the continental margin, and full authority over the deep oceanbed, will depend largely on the portion of the revenue to be allotted to international purposes, as well as on the stringency of the conditions imposed on the licensees in this intermediate zone.

Another approach, advocated by Professor Andrassy and the Commission to Study the Organization of Peace in its twenty-first report, "The United Nations and the Bed of the Sea (II)," would redefine the continental shelf by a combination of depth and width limits, beyond which the oceanbed would be outside national jurisdiction. The commission recommends a 200-meter depth or a fifty-mile distance from the coast, whichever gives most to the coastal state. Professor Andrassy advocates a distance of thirty miles from the coast, and beyond this limit to a depth of 200 meters. Professor Henkin argues for a 200-meter isobath, with a minimum shelf for all nations of a determined number of miles from the coast, together with a buffer zone, in which coastal states would have mining privileges.[4] This last proposal was adopted with some modifications by the United States Draft Convention.

Whatever the precise delimitations or combinations between depth and distance, the essential feature of all these proposals is their strict definition of the limits of national jurisdiction, without which no ordering of the complex problems of the oceanbed is possible.

The opposition to international controls centers on the assertion that national interests must prevail over international utopias, and that it would be foolhardy for countries such as the United States to forego any exclusive rights over portions of the oceanbed claimed to be within the national jurisdiction for the sake of the world community. The opposition will become louder as political crises threaten the supply of oil from the Middle East, or nationalized foreign-owned mining interests in Latin America.

Such arguments have short-term appeal but they are greatly outweighed by the international chaos that will be an inevitable consequence of present trends. In the first place, there are contradictions even within the different countries. In the United States, for example, the State Department's quest for some degree of international order is opposed by the Department of the Interior, which favors the claims of national oil and other industrial interests, but it is strongly supported by the Defense Department, which, in the tradition of the major naval powers, encourages a narrow national jurisdiction for the sake of maximum mobility on the oceans. Closely linked with the first is the second argument that the extension of national jurisdiction is a game that two—and more than two—can play. The expansion of national claims by one coastal power will inevitably be met by equal or greater claims of other coastal powers—including potentially hostile ones. In many cases the disadvantages of greatly restricted access to the seas claimed to be within the national jurisdiction of another state will more than outweigh the advantages gained by an extension of national jurisdiction.

Nor is this whole issue simply a matter of conflict between developed and the developing countries. Landlocked states generally favor an international regime which would give them some share in the resources of the sea. Those economically undeveloped states with a rich land mass extending under the sea might tend to favor a wider national jurisdiction. But since many of them are unable to exploit these resources on their own, they will in fact prefer an international regime to the alternative of total technological, economic, and ultimately political dependence on one of the major industrial powers. The largely poor and undeveloped Latin American states now almost universally claim a 200-mile zone of territorial waters. This is not only to reserve fisheries and other resources for their exclusive national use; but, as we have seen, in some cases it is based on the fear that the exploitation of seabed mineral resources may jeopardize the profitability of land-based mining which is a vital source of income.

The result of all these contradictory and conflicting claims can only be a general increase in international tension and confrontation. Even the most short-sighted advocates of "national interests" can hardly welcome a world in which groups of states will claim vast stretches of the seas around them as their own, while others extend seabed operations further and further outward, with the inevitable result of increasing curtailment of international fishing and navigation, and the threat of confrontation, at the bottom of the oceans. Today's "realism" becomes the madness of tomorrow.

---

[1] *United Nations Draft Convention on the International Seabed Area*, Article 27, pp. 14-15, August 3, 1970.

[2] *Op. cit.*, p. 3.

[3] Frank LaQue, "Deep Ocean Mining: Prospects and Anticipated Short-Term Benefits," *Pacem in Maribus: Ocean Enterprises* (1970), p. 22.

[4] Louis Henkin, *Law for the Sea's Mineral Resources* (1968), pp. 44ff., 73.

7

Structure and Functions of an International Oceanbed Regime

THE major questions concerning the structure and powers of an international oceanbed authority can be summed up as follows:

First, what should be the basis of its constitutional authority? Specifically, to what extent, if at all, should an international oceanbed authority be linked with the United Nations?

Second, what should be its function? In particular, should it be an operative agency, directly concerned with the exploitation of the resources of the seabed? Should it be a licensing authority? Or should it be a purely supervisory and consultative agency?

Third, to what extent, if any, should a new international oceanbed authority become a new development-aid institution, helping in the redistribution of the resources of the earth for the benefit of developing countries and, in view of the new disparities created by the structure of the oceanbed, reallocating a portion of the revenues from oceanbed exploitation to landlocked and other countries disadvantaged by their geographical situation and geological structure?

On the first question, the weight of expert opinion is strongly in favor of an agency linked with the United Nations—although not necessarily limited to its members—but not under direct United Nations control, and structured somewhat differently from other United Nations agencies. The reasons for this attitude are evident. As the United Nations has expanded in membership, it has become imbalanced because of the decline of the Security Council's authority and the corresponding increase in the relative weight of the General Assembly, in which all members have equal votes. Nominal voting power in the General Assembly is no longer relevant to political power, financial responsibility, and technological capacity. Resolutions tend to be determined more and more by political bloc alignments rather than by practical considerations and needs. Far and away the most effective international agencies affiliated with the United Nations are those like the World Bank and the International Monetary Fund, which are operationally and functionally autonomous. Moreover, the effectiveness of an international oceanbed regime would be proportionate to its universality; the most glaring deficiency of an international oceanbed authority strictly linked to the United Nations would be the absence of Communist China. While it is by no means certain that China would adhere to any international agency, it is essential that the possibility be left open. An organization permanently excluding one of the three leading world powers would be gravely handicapped from the start.

International Machinery—
The Secretary-General's Report

On May 26, 1970, at the request of the United Nations General Assembly, the Secretary-General published a "Study on International Machinery," which reflects the rather diverse views of the member states. It distinguishes four basic types of international ocean control:

1. International machinery for exchange of information and preparation of studies.
2. International machinery with intermediate powers.
3. International machinery for registration and licensing.
4. International machinery having comprehensive powers.

The first type of organization would limit an international agency strictly to dissemination of information and the preparation of studies. While one should not underestimate the usefulness of this type of international agency of which the Intergovernmental Oceanographic Commission (IOC) and the Intergovernmental Maritime Consultative Organization (IMCO) are examples, an organization with these functions alone represents a virtual abandoning of any effective international control and an almost total acceptance of a national grab race for control of the oceanbed.

The second type would have slightly more extended functions, such as the preparation of resolutions, the encouragement of scientific research, and the preparation of conventions and international regulations. "The machinery would not, in principle, itself have direct powers but would provide a means whereby states could discuss the issues and adopt certain common solutions, as well as receiving assistance on some of the technical questions involved." This advisory and recommendatory rather than executive kind of organization is comparable to the functions of most of the United Nations specialized agencies. However, even this slightly stronger type of international agency would lack the authority and the power to stop the continuing partition of the oceanbed.

The third proposal would confer the functions of registration and licensing upon an international authority, with separate legal status.

The fourth type of international machinery occupies much the longest part of the report. It would have two types of function which, in this writer's opinion, should be more clearly distinguished: licensing and direct operational involvement. This kind of comprehensive regulatory type of international authority poses certain problems. First, there is the question of the scope of authority. Should an international seabed regime have control over living as well as mineral resources? Clearly, the two are and will be increasingly closely linked. Growing mineral exploration and exploitation will more and more affect the ecology of the oceans, and particularly the benthonic type of living organism. Drilling operations, oil leaks and other forms of pollution, thermal changes, and many other aspects of mining have a direct impact on fishing. There are only two alternatives: either an international oceanbed regime with authority both in regard to mineral and living resources, or close co-ordination between an international oceanbed regime for minerals, and an international fisheries organization—neither of which exists at this time.

There is also the question of what type of organization should be held qualified to apply for a license. Within the continental shelf zone, and also within the international trusteeship zone proposed by the United States draft convention, obviously only the coastal state concerned can have the authority to grant licenses to public or private enterprises, or to consortia, as the case may be. There has been some debate whether only states or other groups as well should be entitled to apply for a license in the international zone proper—that area which would be under the unrestricted control of the international seabed authority. The United States draft permits any kind of enterprise to apply for exploration or exploitation licenses from the authority, if they are sponsored by one of the member states to the convention. This would permit the states to screen the qualifications of the applying enterprise—a necessary safeguard, and one that could hardly be left in the hands of the international seabed authority itself. An alternative, and perhaps a simpler solution, would be to confine applications to member states and leave it to the latter to sublicense enterprises within their jurisdiction. This would clearly place the administrative and legal responsibility upon the member states, which is probaby a more practical solution in the present state of international society. As we have seen, recent British and French government proposals favor the exclusive licensing of states, or groups of states, which would sub-license enterprises and vouch for their technical and financial competence. If corporations are permitted to apply directly, they might be government-controlled (as they would inevitably be in socialist systems) or privately owned. It would therefore be quite essential that there be no doubt about the exclusion of a privileged status for government corporations as, for example, with regard to jurisdictional immunities.

In the evolution of international law an interesting question will be that of the substantive law to be applied to international seabed operations. This is beyond the jurisdiction of any one state, and obviously no particular national law can be applied as such. This will be another potential area for the application of the so-called "general principles of law recognized by civilized nations." In other words, in the solution of any disputes, the tribunal proposed in the United States draft, or any other judicial authority, would have to apply the general principles of contract law culled from the leading legal systems of the world, and particularly of concession agreements concerning the exploitation of natural resources that have been made mostly since the last world war.

One of the most crucial questions would be the elaboration of an equitable allocation of licenses. The Secretary-General's report mentions as alternative criteria, "a first come first served basis, the drawing of lots, grants on the basis of the merits of the applicants, and competitive bidding." It also stresses that "consideration would have to be given to the needs of developing nations, bearing in mind the exploitation of seabed resources for the benefit of mankind as a whole." All this, of course, will be one of the principal objects of negotiation in establishing the terms of the seabed treaty. Foremost among the considerations must be not only the equitable distribution and principles of nondiscrimination but also the prevention of overcrowding, which would increase pollution dangers, and other threats to ocean ecology.

We will now compare the Secretary-General's survey with some of the most important official and nonofficial proposals for the organization of an international oceanbed regime.

The most common feature of the various proposals is an authority open but not limited to members of the United Nations, organized—like most of the specialized agencies of the United Nations—into a three-tier structure of a general assembly, a council, and an executive staff headed by a director general, and differing from the United Nations in that representation and voting are weighed according to the financial and operational contributions made by the various members. Thus the plan of the Commission to Study the Organization of Peace proposes a United Nations seabed authority, whose organs are the sea assembly, the sea council, and the director general. The assembly has "contributing" and "regular" members, the former consisting of those members of the United Nations that contribute more than a certain percentage to the general budget of the United Nations. The sea council will be composed of contributing members and six regular members, four from coastal and two from noncoastal states. The council will be authorized to issue licenses, collect fees, prevent pollution, and generally control the conditions of exploitation of seabed resources. As in the case of other international agencies, it is the director general and his staff who will carry the main burden of executive operations.

The plan put forward by the Center for the Study of Democratic Institutions would establish a maritime commission and a maritime assembly. The former would be composed of seventeen members serving for three years, including the five member states most advanced in open space technology. The others would be elected by the maritime assembly, with due regard to adequate representation of developed and developing, maritime and landlocked, and free-enterprise and socialist countries. The maritime assembly would be divided into four chambers. The first would be elected by the General Assembly of the United Nations, divided into nine regions of the world. The second would represent international mining corporations, producers, consumers, and others directly interested in the extraction of seabed resources. The third would represent fishing organizations of all kinds—processors, merchants, seamen's unions, and fishery commissions. The fourth would represent the various scientists concerned with the oceanbed (oceanographers, maritime biologists, meteorologists, and others).

A draft treaty sponsored by Senator Claiborne Pell proposes that a specialized United Nations agency be made the licensing authority over deep-sea operations. The licenses would be issued to governments, which would act as entrepreneurs themselves or issue sublicenses to public or private enterprises within their jurisdiction.

The United States Draft Convention of August, 1970, is the most recent and influential of the organizational proposals to the United Nations Seabed Committee. It suggests an international seabed resource authority composed of an assembly, a council, and a tribunal. All contracting parties would make up the assembly whose most important function would be approval of the budget proposed by the council, and of changes in the allocation of net income. The council would consist of twenty-four members, including the six most industrially advanced contracting states, at least twelve developing countries, and at least two landlocked or shelf-locked states. Decisions made by the council would require the approval of the majority of both the six most advanced industrial members and the eighteen other members. The council would appoint the secretary-general of the authority, prepare the budget, and establish procedures for co-ordination between the authority, the United Nations and other relevant international organizations. Together with the secretary-general and his staff it would clearly bear the chief administrative burden.

The function of the tribunal under the proposal would be most important. It would have compulsory jurisdiction to decide all disputes between the contracting parties, advise on all questions relating to the interpretation and application of the convention, have power to impose fines, as well as to award damages, and, in the case of gross and persistent violations of the provisions of the convention, to either revoke the license or request the trustee party (in the intermediate zone) to do so. Where the contracting party failed to perform the obligation imposed upon it by a judgment of the tribunal, the council would enforce the judgment by suspending the rights of the defaulting party.

The most important subsidiary organs under the proposal are a "Rules and Recommended Practices Commission," an "International Seabed Boundary Commission," and, above all, an "Operations Committee," which would issue licenses for mineral and other exploitation in the oceanbed beyond the international trusteeship area, as well as supervise the operations of licensees, in co-operation with the trustee.

The proposal to give special representation to the nations most advanced in ocean technology parallels the constitution of the International Atomic Energy Authority, which grants a privileged status to the countries most advanced in nuclear technology. This necessary departure from total nominal equality is balanced by adequate representations of nations that are disadvantaged in terms of development and geographical situations.

The United States draft supports the almost unanimous opinion that the seabed authority should not be operational. There is general agreement that even a seabed authority detached from the cumbersome mechanism of the United Nations itself would be ill-equipped to undertake the investment and the complex technological operations involved in the exploitation of the seabed resources.

Of great importance is the question whether the international seabed authority should become a development-aid agency, since development aid, a burden upon national budgets and administered by public national-aid-development agencies is in a state of deep crisis. In fact, it is dying in the United States, which was by far the most important single provider of development aid in the postwar period. While the Pearson Commission Report on "Partnership in Development" (1969) and other authoritative official and unofficial studies have postulated 1 per cent of the G.N.P. as a minimum ob-

jective for development aid by the richer nations, the
United States' percentage has declined to .4 per cent
and is likely to fall further. At the same time, the absolute and relative needs of developing countries increase
all the time. The number of applicants grows constantly,
as more and more small and poor nations become independent. Many countries find that the burden of repay-
ing the debts of loans received earlier absorbs most of any
further development aid they receive. In addition, the
poorer and economically underdeveloped nations are
also by and large those that depend primarily on the ex-
port of stable commodities such as coffee, tin, rubber,
copper, cocoa, or palm oil. While the prices of industrial
products rise all the time, the world market prices of
primary commodities have either remained stationary or
actually declined ever since the end of the Korean War,
widening the gap between rich and poor nations. The
weight of international development aid is shifting to
the international agencies devoted to this purpose, and
particularly to the World Bank and its affiliates, which
under the presidency of Robert McNamara have greatly
increased the volume of its lending. But although
the World Bank and some of the regional development banks, such as the Inter-American Development
Bank, have capital resources of their own supplied
by the quota contributions of their members, and although some of them are able to borrow additional resources on the world money market through bond issues, their capacity is limited. There is an urgent need
for new sources of development-aid capital. In principle, the idea of allotting a share of the oceanbed revenues to development purposes, as espoused by the Pardo
plan and the United States proposal, is therefore to be
welcomed. But unless the authority receives revenues

from areas in addition to the abyssal depths, its income will be negligible.

All this emphasizes the inevitable interrelationship of the ocean's various uses. The worst of the many possible future prospects is that of almost unmitigated international chaos on and below the seas. Not only would coastal states continue to expand their national jurisdiction, granting more and more licenses without any agreed international standard, but oceanbed activities would not be co-ordinated with shipping, fishing, scientific research, cable and telecommunications, and other uses of the seas. It would be a free-for-all fight between state and state and between one use of the sea and another. At the other end of the spectrum we may envision a fully co-ordinated international sea regime, in which the concerns for the exploitation of mineral resources are co-ordinated with those of marine biology, shipping, pollution prevention, and cable laying, and freedom of shipping, fishing, research, and exploration are preserved as "the common heritage of mankind."

8

Ocean Enterprise— the Organization of Oceanbed Exploitation

THE exploration and exploitation of growing portions of the oceanbed is already a fact—and one of growing international, economic, technological, and political importance. In the United States offshore oil constitutes a growing proportion of the nation's total production. Virtually all the major multinational enterprises engaged in the exploitation of oil, gas, nickel, and other mineral resources have obtained leases and licenses. Countries poor in mineral resources, such as Britain, have divided up their continental-shelf areas between a variety of enterprises of various nationalities, and in some cases state-owned corporations, such as the British Gas Council, have formed joint ventures with the licensees.

It is most important that the skills, capital, and interests of the various state or privately owned enterprises and of multinational consortia should be co-ordinated with the interests of the international community, as they hopefully will be represented by an international seabed regime. This means the development of licensing and enterprise organizations that will protect both the interests of the enterprises involved and those of the various states concerned with the exploitation of oceanbed resources, particularly the developing countries. Moreover, entrepreneurial activities must be linked with conservation in the exploitation of the oceanbed. An official inquiry into the causes of the highly damaging oil leaks from operations off the coast of Santa Barbara revealed that the United States federal safety standards had been violated in hundreds of cases even by the leading companies. An interrelationship between licenses, entrepreneurial standards, and regulatory controls is therefore vital. In this respect, the role of insurance will become increasingly important. As a leading British insurance expert has pointed out, "safety under the sea has been and will hopefully continue to be regulated by insurance, working hand in hand with the oceanographic industry's own rigorous, self-imposed safety guide lines. Insurers as well as oil men and aquanauts hope for sensible, workable legislation from international lawyers who have a clear knowledge of whom they are protecting and from what."[1]

The task is to devise forms of ocean enterprise that will provide sufficient incentive for the corporate entrepreneurs of the various states—governmental or private —to undertake these expensive and complex operations while safeguarding the international community and enabling the great majority of countries that cannot be entrepreneurs because of their geographical situation, lack of financial resources and technological know-how, to be actively associated with the exploitation of ocean-bed resources.

We have already dealt with the public regulatory aspects. Clearly, the minimum standards required to provide safety for shipping and fisheries, to enforce adequate pollution controls, and to implement other aspects of international community interests, such as nondiscrimination in licensing procedures, must be incorporated in any concession contract operating outside the zones of national control. Under the United States proposal this would apply to the trusteeship zone as well as to the deep oceanbed that would be under purely international authority. Since the continental shelves will for a long time constitute by far the most important source of exploitation of living and of mineral resources, it will be essential that corresponding standards be adopted by the coastal states even within their national jurisdiction. This could be provided for by a protocol to any international seabed treaty or by separate concurring declarations. Indeed, the United States could further advance its pioneering effort by a unilateral proclamation and incorporation into national licenses of the standards it wishes to see included in an international treaty.

Joint Ocean Ventures

The joint international business venture has become the increasingly dominant form of association between both public and private enterprises of different countries, particularly in the relations between developed and developing countries. It is submitted that this should be the form adopted for ocean enterprises. Binational and multipartite partnership agreements, in particular joint ventures, also afford an excellent opportunity for development-aid institutions—international, regional, and national—to assist financially, and thereby to associate, groups of countries with the development of ocean resources.

There is a great variety of economic systems, and much is made of the contrast between "socialism" as represented by the U.S.S.R., and the United States-style "capitalism." But there are also many intermediate systems, generally described as "mixed economies," in which state and private enterprise coexist, as, for example, in Britain, France, Italy, and India. Since all these states are likely to play major entrepreneurial roles in the exploitation of the oceans, it is important to emphasize that these differences in economic organization have only minor international significance. It matters relatively little whether an enterprise is undertaken by a state or a private corporation. In the case of the Soviet Union or Italy (whose ENI[2] is a dynamic state-chartered corporation) there is a direct link between the enterprise and the government; in Britain and France, oil and gas enterprises have less direct links with the government. Of Britain's two giants, British Petroleum has a substantial government holding but is conducted like a private enterprise, with certain government representatives on the board. Shell, which is entirely privately owned, is a joint British-Dutch enterprise. Both are among the multinational oil giants. In France, ERAP[3] is a government-owned oil enterprise that has entered into a number of international agreements, for example, with Algeria. The Compagnie Française des Pétroles

(CFP) is essentially a private corporation with a minority government holding, somewhat comparable to BP.

In the United States the oil industry is entirely privately owned, but the link between oceanbed activities and government policy and regulation is obvious. From the territorial waters to the edge of the continental margin the rights and jurisdiction claimed by the coastal state are in the hands of the government.[4] Exploration and exploitation licenses for the subsoil of territorial waters, the continental shelf, and any area beyond which the coastal state claims jurisdiction must be given by the appropriate government department. In the United States these are issued by the Department of the Interior, whose policy is often at odds with those of the more internationally oriented Department of State and the Department of Defense. The latter has an interest of its own in freedom of movement under the surface and along the ocean bottom, which is quite often in conflict with the industrially oriented Department of the Interior. Whatever the internal conflicts of government departments and agencies may be, it is ultimately the government that controls the extent of commercial exploration and exploitation on the ocean floor. In matters of armament traffic it is not very important whether the manufacture of arms is operated directly by government enterprise or by private manufacturers and traders. No modern government lacks the power to control arms trade if it wishes to. The many abortive international agreements to control arms traffic, as, for example, the Non-Intervention Agreement during the Spanish Civil War, faltered not because of lack of governmental power to enforce the conventions but because of political considerations. We need not therefore attach too much importance to the difference between capitalist, socialist, and intermediate systems with regard to the control of the oceanbeds. Neither the Soviet Union nor the United States, Britain, France, nor any other country that matters would allow the form of its economy to interfere with its participation in the present race to the bottom of the seas.

As in the case of many human endeavors the keynote of ocean enterprise must be variety and flexibility. This applies to the area as well as to the form of international co-operation. How, then, can the interests and skills of public or private mining entrepreneurs be combined with the needs of the world community?

First, there are certain situations in which the close geographical proximity of various coastal states and their joint interest in the exploitation of a certain maritime area will make it desirable for two or more governments—or their chosen agencies—to co-operate in a joint exploration venture. Such areas include the Gulf of Mexico, the Caribbean Sea, and the Persian Gulf. Up to now, however, the record of exploitation in these areas has been partition rather than co-operation. There have been no joint ventures between the United States and Mexico—whose coastlines enclose the Gulf. Instead, the coastal states and in particular the United States have issued separate licenses within their national jurisdictions. In its judgment in the North Sea Continental-Shelf Case the International Court of Justice observed that the North Sea forms a natural unity. Since its depth nowhere exceeds 200 meters, except for the so-called Norwegian Trough, it is well within the universally accepted depth of the continental shelf. The coastal states—Great Britain, France (if the Channel is included), West Germany, Belgium, The Netherlands, Denmark, Norway—have many political and commer-

ment of the oceans. Such a corporation, moreover, would remove any stigma that might attach to a private operation."[5] The only doubtful aspect of this observation is the term "corporate sovereignty." This clearly must rest with an international oceanbed authority formed by the community of nations and equipped with licensing and police power.

Over the past fifteen years a variety of bipartite, multipartite, intergovernmental, private, and mixed public-private enterprises have been set up to pool both the resources and skills of different countries in matters of joint concern. A few examples will illustrate the experiences from which international seabed exploitation could profit.

In 1955, sixteen European continental countries set up a European company (EUROFIMA) that financed railway equipment for the purpose of unifying and improving the construction and performance of railway rolling stock. Although established as a Swiss company —in the absence of an international company law—it is an international corporation, the capital of which is subscribed by the various national railway administrations. A similar company (EUROCHEMIC), set up in 1957 for the chemical processing of irradiated fuels, and incorporated in Belgium, has as its members not only governments and public authorities but also public, mixed public-private, and private corporations. France, West Germany, and Luxembourg set up the International Moselle Company for the construction of the Moselle Canal. In 1953 the French and Italian Governments jointly established a company for the construction of the Mont Blanc Tunnel, the operation of which after completion was taken over by a joint stock company, itself composed of a French and an Italian company.

Air Afrique was formed in 1961 by eleven states, formerly French-African territories, in association with a French company, to operate air services in that area.

The International Telecommunications Satellite Consortium (INTELSAT) is a different type of transnational organization, the manager and major shareholder of which is a private, though government-supervised, United States corporation (COMSAT). The other members are a large number of states represented by their respective postal and telephone administrations or other public organizations. A joint venture between one or several private corporations of technologically advanced countries and a developing country represents a particularly important form of association that could be utilized by developing countries for oceanbed exploration. In cases of major capital investment such ventures are often supported by loans or equity investments from international, regional, or national development agencies.

It may suffice to give two typical illustrations from the recent history of the development of land resources, both of which are applicable to the development of seabed resources. One is a bipartite venture between a developing country and an overseas investor. The other is a multipartite venture in which the government of a developing country, a multinational consortium of private foreign investors, and various public and private financial agencies participate.

The Chilean and Zambian copper mines, among the most important in the world, used to be both owned and operated by foreign companies. The major foreign investors in Chile are the Anaconda and Kennecott Corporations; in Zambia it is a consortium of British and American companies. In both cases the governments have converted the foreign holdings into 49 per cent participation in the past few years by partial nationalization measures. The government, directly or through a government-owned corportion, has taken over a 51 per cent majority of the stock, leaving the management and operation in the hands of the foreign investor while a native staff is being trained.[6]

The LAMCO venture in Liberia is a $300 million project concerned with the exploitation and marketing of high-grade iron ore produced in the Nimba Mountain several hundred miles from the coast. Under a complex multipartite arrangement, first set up in 1960, a fifty-fifty joint equity venture was established between the government of Liberia and a foreign consortium, with a small capital of a few million dollars. The government of Liberia paid for its share through a concession to mine the iron ore on its territory. It has, as it were, brought in its sovereignty over the exploitation of natural resources. Built around this joint equity venture are much more complex capital investments and loan arrangements. The foreign consortium is itself a joint venture between a number of mainly Swedish companies, which also supplied a large amount of the loan capital in addition to its equity share. A 25 per cent share of this investment was taken over by the Bethlehem Steel Corporation in return for commitments on the delivery of iron ore to it. Long-term sales contracts have been concluded with foreign buyers, mainly in the West German Ruhr area. The principal German public-aid agency—Kreditanstalt für Wiederaufbau—and the United States Export-Import Bank have both made substantial loans for the capitalization of this huge venture, which involves not only the mining of iron ore but the construction of rail and port facilities and attendant social services, such as schools and hospitals.

Joint exploitation need not necessarily be in the form of joint equity ventures. In the extraction of oil, for example, contractual joint-venture arrangements have been more frequent. There is a precedent for such a joint structure in offshore exploration in certain parts of the Persian Gulf. In 1965 the government-owned National Iranian Oil Corporation entered into an offshore exploration and exploitation agreement with three foreign parties. The NIOC participates with 50 per cent of the capital (and profits). The other 50 per cent is shared in equal parts between AGIP, the subsidiary of the Italian state-owned Ente Nazionale Idrocarburi; the state-owned Indian Oil and Natural Gas Commission (ONGC); and Phillips, a private American oil corporation. The parties to this agreement formed a joint managing company called IMINOCO, which is registered in Iran.

In many cases of land-based mining and other concessionary operations in recent years, a major problem has been the nationalization of foreign-owned operations by the state in whose territory the resources are located. The amount and manner of compensation in such cases has been the subject of much political friction and legal controversy. These problems would not arise in the case of seabed operations, which have the advantage of starting with a clean slate.

Another question that will arise in the case of sea mining, as it has for land-mining operations, is the way in which the undeveloped country can pay for its share in a joint operation. Here, the difference will be important between operations within and those outside the limits of national jurisdiction. Within the continental-shelf area, the coastal state now has clear sovereignty over seabed and subsoil resources, whether by convention or custom. In this and any further area of national sovereignty the government concerned could therefore contribute its share, as in the case of LAMCO, by granting to the foreign partner or consortium the right to conduct operations within the area of its sovereignty. Obviously this would not apply to the ocean area outside national jurisdiction and presumably under the direct control of an international seabed authority. Here, the developing country would have to make a certain financial investment; and since the investment in deep-sea mining operations is considerable, it is an area where joint ventures between developing countries—possibly through regional development corporations—would be particularly appropriate. The other party, whether it is a private or public entrepreneur, will undoubtedly insist on a return proportionate to its commitment of capital and skill. In this respect, the necessary adjustments in favor of developing countries may have to be made through the international seabed authority, which will receive a proportion of the revenue in the form of royalties and fees.

The examples given here show that the economically and technologically less-developed countries may use the joint-venture pattern to obtain an appropriate share —and in the case of continental-shelf resources, presumably a majority control—in operations for which they need the capital investment and the technological skill of foreign entrepreneurs. Furthermore, there are many precedents for joint operations between governments and private enterprises, or between state-owned and privately-owned groups, which can bridge the differences of economic organization between socialist and capitalist countries. In areas under the direct control of the international seabed authority it may well be that multinational, intergovernmental ventures, or joint ventures between governments and private enterprises (as in the case of EUROCHEMIC), at least in the beginning, would provide the most suitable form of operation.

Joint Enterprises in Fisheries

Reference has already been made to the close connection between exploitation of the nonliving resources of the ocean bed and the fisheries. While the population explosion will make us increasingly dependent upon more systematic and scientific uses of the living resources of the sea, at the same time, these are gravely threatened by two aspects of modern technological development. One is the threat to marine life posed by explosions, oil pollution, radioactivity, and other activities connected with oceanbed exploration. The other is the threat of reduction, and eventual extinction, of fish and other valuable living resources of the sea through the ruthless and competitive use of mechanized methods of fishing and fish processing. It seems almost inconceivable that at a time when the conservation and scientific management of the living resources of the sea is of growing importance there should be an almost total absence of effective international regulation, let alone co-operation, in the management and distribution of fisheries. But such is the case. Successive international whaling agreements concluded between 1931 and 1946 (the latter establishing a permanent International Whaling Commission) have not prevented ruthless overkill by the world's major whaling fleets, equipped with modern electric harpoons and factory ships, and the virtual extinction of the blue and humpback whale.

The chairman of the Committee on Whales for the Environmental Defense Fund pointed out in a letter to the *New York Times* of August 21, 1970, that Japan and the Soviet Union today account for 85 per cent of the world's whaling, although "five nations were participating in the plunder of the Antarctic fishery." As the major importer of whale oil for a variety of purposes including luxury consumption, the United States has contributed to the plunder by implicitly encouraging the continued operations of whalers, even at a time when the whale-oil yield has dropped from 2,000,000 barrels to fewer than 400,000 barrels within a decade. The ban on the import of whale products issued by the United States Department of Interior under the Endangered Species Act of 1969, may help to prevent total disaster. Perhaps the imminent killing of the goose that lays the golden egg may induce the major whaling states at last to make effective the International Whaling Convention and the functions of the International Whaling Commission. But without close and continuous international inspection on sea and on land there is little hope. The Geneva Conference of 1958 adopted a convention on fishing and conservation that has so far been ratified by only thirty states of the eighty-six states represented. Its principal provision allows a coastal state with "a special interest in the maintenance of the productivity of the living resources in any area of the high seas adjacent to its territorial sea" to take certain conservation measures. The convention also imposes the duty on states whose nationals fish in a particular area of the high seas—where the nationals of other states are not so engaged—to adopt conservation regulations for that area.

Since the overwhelming majority of exploitable pelagic, demersal, and benthonic fisheries resources and their breeding grounds are within the continental-shelf zones, frictions are increasing because of the likelihood of increased conflict between the coastal states in respect to jurisdiction over the seabed and subsoil resources of this area, and the subsisting international rights of fishing in the waters above. It is perfectly comprehensible that in the absence of effective international controls coastal states should seek to take measures to preserve fisheries and breeding grounds within their adjacent waters. So far only the states that have proclaimed 200-mile territorial zones have openly excluded foreign fishing fleets from these zones. But, inevitably, other states will seek to extend their exclusive jurisdiction from the bottom of the continental shelves to the surface. This is an understandable reaction in the absence of international management of fishery resources and in view of the ruthlessness of some nations in overfishing in the continental-shelf areas of other nations. It is an open secret that Russian trawler fleets with the latest equipment fish in the continental-shelf zone of the Canadian coast, where they move from one area to another. All attempts to reach a bilateral agreement have failed. Canada reacted recently with the proclamation of a twelve-mile territorial water limit, measured from base lines that in effect extend the zone to roughly one-third of the continental shelf, and the assumption of power to declare exclusive fishery rights in the 100-mile "environmental protection" zone in the Arctic. Since trawlers of all nations can easily be equipped with military-intelligence devices, the tensions in the continental-shelf area will almost certainly increase.

The need for an effective international fisheries organization and joint enterprises in the exploitation of the living resources of the seas is perhaps even stronger than in the field of mineral resources. There has been much talk of "fish farming"—more systematic methods of breeding, feeding, and farming of fish than those used now. The possibilities of systematic sea farming are illustrated by such examples as the production of kelp (seaweed) off the coast of California at an annual rate of 160,000 tons, and Japanese oyster farming, which has raised the yield of oysters from 600 pounds to 32,000 pounds a year. The United States Public Health Service has designated 10 million acres of sea as suitable for shellfish farming. At a rate of 600 pounds per acre, 6 billion pounds of shellfish could be produced annually, a figure equal to the entire United States fish catch.[7] But fish farming must be undertaken in co-ordination with conservation; excessive exploitations by any one coastal state without regard for the ecology of the ocean at large might lead to the same disastrous consequences that have already occurred in whaling.

Obviously, international scientific collaboration, international licensing procedures, and joint ventures in the exploitation of fisheries will have to go hand in hand. Methods and procedures should not differ vitally from those already analyzed for the exploitation of oceanbed resources. While there will be exclusive rights for fisheries within at least the twelve-mile territorial water limits, it is highly desirable that international breeding and conservation standards should be agreed upon by all the maritime states, even for their exclusive national zones. Beyond the "limits of national jurisdiction," an international fisheries authority should be empowered to issue licenses for the exploitation of fish and other living organisms in defined areas. The actual exploitation could be undertaken by joint ventures between different nations, very much in the manner of industrial enterprises. For it is clear that fishing will become increasingly industrialized. The process could parallel the way in which the ruthless deforestation of vast areas of the continent, particularly in North America, has gradually given way to forestry management, contour plowing, and soil ecology. The type of partnership association in fishery ventures would differ from that of industrial mining in that the enterprise leaders in fisheries and fishing equipment are nations such as Japan, France, or Norway. Also, the capital investment is likely to be smaller than in industrial mining corporations.

It should be apparent from the examples given that there is a great variety of possibilities available for exploitation of the oceanbed resources, which would take into account the enormous diversity of present and future interests and situations. Legal devices or organizational problems are not the major obstacles, but the reluctance to make co-operative arrangements and to appropriately apportion the wealth of the oceanbed between the privileged and the underprivileged. In the postwar period such adjustments have gradually occurred in land-based operations. Former colonies or dependencies have acquired sovereignty and with it the claim to control their natural resources, as it was expressed, for example, in the United Nations Resolutions on Permanent Sovereignty over National Resources, in December, 1962. At the same time, the enterprises of the

capital-exporting and industrially developed countries
—which in previous decades obtained concessions more
or less at will and were often able to dictate their terms
—have come to accept the legitimacy and inevitability
of the claims of the new countries. It would be absurd
and anachronistic to start on a different assumption in
the exploration of seabed resources.

A Summary of the Organization of Seabed Exploitation

The type of organization for the exploitation of
seabed resources will generally differ inside and outside
the limits of national jurisdiction. Within these limits,
the coastal state will normally license entrepreneurs at
its own discretion. In technologically and industrially
advanced countries most of these will be nationals of
the coastal state. But the example of Great Britain
shows that licenses may be granted to bidders of different nationalities on competitive terms. The license will
either be in the form of pure concession—an operation
in which the government owns the entire equity and
the operator receives a fee—or more frequently the reverse—a concession in which the entrepreneur owns the
equity interest and pays a royalty on the revenue. There
may be intermediate types of joint ventures between
the coastal government and the concessionaire, which
will often be an international consortium.

In the case of undeveloped countries, continental-
shelf exploitation more frequently will be a joint venture, perhaps of the type of the Iranian offshore operations. The coastal government in these cases will generally prefer a pure contractual arrangement under

which the foreign entrepreneur would work for a fee. But sufficient interest in such management contracts is unlikely on the part of foreign enterprises, which generally will prefer an equity share.

Outside the limits of national jurisdiction—as a minimum, for the oceanbed beyond the continental margin, and as a maximum, for the oceanbed area beyond the redefined continental shelf—control over operating conditions would be in the hands of an international seabed authority. Subject to the terms of an international treaty, this authority could lay down the conditions of the concessions granted, including joint ventures if any, the shares of the partners, the royalties to be paid to the international authority, and the other conditions of operation. If it were adopted by international treaty, in the intermediate zone proposed by the United States Draft Convention, the licensing, including the determination of the type and form of joint ventures, would be in the hands of the coastal state, subject to regulatory conditions supervised by the international seabed authority, which would also determine the share of the profits to be paid into an international pool.

---

[1] James W. Dawson, "Insurance as a Regulator," in *Pacem in Maribus: Ocean Enterprises*, p. 38.

[2] Ente Nazionale Idrocarburi.

[3] Entreprise de Recherches et d'Activités Pétrolières.

[4] In the United States there have been a number of disputes as to the respective rights of the federal and state governments—such as California, Texas, or Louisiana—with regard to sovereignty over the adjoining waters and continental shelves.

[5] Richard Eells, "Emergence of a Corporate Sovereignty for the Ocean Seas," *Pacem in Maribus: Ocean Enterprises*, p. 64.

[6] As this book goes to press, it appears likely that the new Allende government of Chile will fully nationalize all foreign holdings in mines.

[7] Lawrence Galton, "Aquaculture Is More Than a Dream," *New York Times Magazine*, June 10, 1967, p. 13.

9

The Future of The
Oceans – the Choice
Before Us

EVEN when taken by itself, the question of whether the oceanbed and inevitably the seas themselves will be partitioned, or whether this area which represents seven-tenths of our globe will become a field of international co-operation, is of immense importance. But it has even wider implications; it is a vital aspect of the choice that faces the world in every field. Will mankind divide itself more and more into contending blocks, which at best will keep an uneasy peace based on a precarious balance of power and at worst become involved in an all-destructive war? Or will the political conscience of contemporary man awaken in time to realize that intensive and worldwide co-operation in all spheres—security from aggression, population growth, conservation and development of resources, communication and transport, minimum standards of human welfare—is a matter not of idealism but of survival?

Although the record of the postwar period is disappointing, it is not one of unmitigated gloom. Nations have begun to co-operate in important areas. Many new institutions have been created that are at least a token realization of the inescapable needs of our time. The host of specialized agencies created within the framework of the United Nations indicate at least the aspiration toward international co-operation in such matters as food and agriculture (FAO), world health (WHO), control of nuclear energy (IAEA), labor and social welfare (ILO), air transport (ICAO), global communications (ITU, IMCO, INTELSAT), and cultural co-operation (UNESCO). But these agencies are all essentially advisory and consultative. With minor exceptions, they have no power to lay down laws of conduct in their respective spheres for the member states. Any resolution, convention, recommendation—even when passed by the appropriate majorities in the international agencies—needs separate acceptance and ratification by each of the signatories.

On a worldwide level of international organization the most significant breakthrough can be seen in the two agencies concerned with international monetary co-operation and development assistance—the International Monetary Fund and the International Bank for Reconstruction and Development (World Bank). Both were established by the Bretton Woods Conference of 1944. Before then, as was demonstrated in the interwar period, the financial crisis of any one country could throw the world into chaos, while economic development was a matter of private commercial exploitation, not of public international responsibility. For all its deficiencies, the International Monetary Fund has been able to regulate the movements between the world's currencies outside the Communist bloc, which has a controlled system of its own, and by and large it has helped to avoid international financial chaos. The World Bank, with its two affiliates—the International Development Association and the International Finance Corporation—symbolizes the new concern for the economic development of the poorer nations, most of which are former colonies. These efforts are now supplemented by a number of regional institutions. Financial autonomy and managerial expertise have combined to detach both the IMF and IBRD from the political maneuvering and conflicts of the general United Nations organization. Both the World Bank and the International Monetary Fund are more comparable in the method of their financing to a shareholding company than to the typical United Nations agency, with their funds supplied by capital subscriptions by the member states, according to their economic possibilities. Although the United States is the largest single contributor to both organizations, their constitutions regulate the voting power in such a way that no single state can obtain a majority. The main advantage of this form of financing is that neither organization has to submit an annual budget for approval to the United Nations, which in turn depends on the contributions of its member states.[1] The IBRD has never had to call on more than a fraction of the member states' subscriptions; it has received a steadily increasing revenue from the interest and service commissions on its loan and technical advisory operations, and it has also been able to raise much additional capital by the issue of bonds, which are highly rated international securities. An equally important consequence of financial autonomy has been the ability of the IBRD and the IMF to appoint highly qualified staffs chosen for ability and expertise rather than on a nationality-quota basis—which is the case in most other United Nations agencies.

It must be hoped that the lessons of the postwar international organizations can be applied to the establishment of an international seabed authority. It will need independence, expertise, and initiative to fulfill its functions, and this means independence from the inevitable multiple political pressures of a direct affiliation with the United Nations.

Some of the most significant developments in international co-operation have occurred within smaller groups of nations. The European Economic Community, formed by six highly developed states that have been at war with each other for centuries, represents an as yet uncertain and incomplete attempt to integrate the economies, the movement of goods and persons, and ultimately the general policies of these countries. This will inevitably mean the strengthening of the supranational aspects of the community and notably the decision-making powers of the Commission, its permanent executive.

The uncertainty of whether and how rapidly these developments will occur, even within the relatively closely knit group of the six, indicates the immense difficulties that lie in the way of an effective supranational authority over the oceans. In one vital respect, the establishment of effective international control over the living and nonliving resources of the oceans could be much easier than the attempt to establish supranational powers in the European community, let alone among a wider group of nations. Both the latter have to overcome centuries of tradition of national sovereignty, and their governments have to surrender some powers for the sake of common purposes. But the seas have been free of national jurisdiction, and exploitation of the oceanbed is a new concern. It is much easier to build up an international regime from scratch than to transform an established system. But for the same reason, every year that passes without effective measures diminishes the prospect of international ocean control. For with every year countries will repeat and extend the pattern they have developed on land: establishing exclusive national, political, military, and economic interests that will be enormously more difficult to modify or abolish than if they had never existed. Time, therefore, is desperately short.

Two immensely destructive world wars have not sufficed to overcome the basic supremacy of the national state. For all its achievements in certain spheres of functional international co-operation, the United Nations has failed in its essential objective: the creation of an effective system of international security that would, at least in major international crises, displace the traditional prerogatives of each state to declare war. The Security Council on the whole has been paralyzed by dissension among its major members, and the General Assembly has become more and more a debating chamber while essential political decisions are taken outside the United Nations. Of the three major world powers, one is not even a member of the United Nations. The United States and the Soviet Union have been able to join forces occasionally when their interests happen to coincide, as, for example, in their joint action in 1956 to stop the British-French-Israeli operations in the Suez Canal Zone, their joint sponsorship of a treaty to ban the proliferation of nuclear weapons, and their agreement in 1963 to stop nuclear testing above ground. At the present time, they also concur on a limited agreement to ban stationary nuclear weapons on the sea floor —a very limited and inadequate move toward the prohibition of military uses of the oceanbed. But an enormously larger area remains where the United States and the U.S.S.R. have not reached even limited agreement. At best there is a standoff, a reluctance to let conflicts go to the point of no return. But, then, events often go beyond the control of the leaders of even the mightiest nations.

All this has tremendous significance for the future of the oceanbed, which in turn threatens to become another major aspect of the worldwide power conflict. In his grim vision of *1984,* George Orwell foresaw three major empires, Oceania, Eurasia, and East Asia, which would absorb the numerous smaller states and live in an uneasy and shifting coexistence. It is possible that an integrated Western Europe and a rapidly rising Japan may have to be added to the small elite of superpowers. The confrontation now threatens to extend to the bottom of the oceans long before 1984. As economic investment and appropriations inevitably entail political control, a confrontation of the powers will take place on land, in the air, and at the bottom of the sea, and it would be absurd to believe that more than 130 nations, or even a small fraction of them, could preserve independence in such a showdown. As we have pointed out, the high degree of scientific and technological knowledge, enormous capital investment, and attendant political and military protection required for the exploitation of oceanbed resources can be achieved only by a very few nations. Even if smaller nations, including the landlocked states, were permitted to share some of the newly mined wealth, they would do so essentially as client states, dependent upon the technological power and military protection of the few superpowers. The only alternative for smaller nations is their closer integration in regional groups, which will enable them to pool their political and economic resources and thus take their place in an international seabed regime in this way. There is no lack of proposals to provide an equilibrium between the various groupings and establish a fair balance between the developed and the less-developed countries, between the big and the small. What is lacking is not the institutional apparatus but the political will. Certainly, we cannot return to a *laissez-faire* world. The stark alternative is between the partition of at least large portions of the oceanbed—and the superjacent waters—and an international welfare regime.

The tragedy of mankind may prove to be the inability to adapt its modes of behavior to the products of its intellect. Twentieth-century man threatens to be a new kind of dinosaur, an animal suffering from a brain ill-adjusted to its environment.

---

[1] How damaging this condition can be for a United Nations agency was illustrated in August, 1970, when the United States Senate, under strong pressure from the bitterly anti-Soviet leadership of the A.F.L.-C.I.O., withheld the United States contribution to the fifty-year-old International Labor Organization in protest against the appointment of a Russian as one of the assistant directors general (the U.S.S.R. has been a member of the I.L.O. since 1954). Because the International Development Agency (IDA) is also financially dependent on periodic contributions from the member states, it has been greatly hampered in its operations by recurrent shortages of funds.

10

United Nations Resolutions of December 17, 1970

On December 17, 1970, the General Assembly of the United Nations adopted—by 108 votes in favor, none against, and 14 abstentions (including the Soviet block)—the following resolution:

Declaration of Principles Governing the Seabed and the Ocean Floor, and the Subsoil Thereof, Beyond the Limits of National Jurisdiction

THE GENERAL ASSEMBLY

RECALLING its resolutions 2340 (XXII) of 18 December 1967, 2467 (XXIII) of 21 December 1968 and 2574 (XXIV) of 15 December 1969, concerning the area to which the title of the items refers,

AFFIRMING that there is an area of the sea-bed and the ocean floor, and the subsoil thereof, beyond the limits of national jurisdiction, the precise limits of which are yet to be determined,

RECOGNIZING that the existing legal regime of the high seas does not provide substantive rules for regulating the exploration of the aforesaid area and the exploitation of its resources,

CONVINCED that the area shall be reserved exclusively for peaceful purposes and that the exploration of the area and the exploitation of its resources shall be carried out for the benefit of mankind as a whole,

BELIEVING IT ESSENTIAL that an international regime applying to the area and its resources and including appropriate international machinery should be established as soon as possible,

BEARING IN MIND that the development and use of the area and its resources shall be undertaken in such a manner as to foster healthy development of the world economy and balanced growth of international trade, and to minimize any adverse economic effects caused by fluctuation of prices of raw materials resulting from such activities,

SOLEMNLY DECLARES THAT

1. The sea-bed and ocean floor, and the subsoil thereof, beyond the limits of national jurisdiction (hereinafter referred to as the area), as well as the resources of the area, are the common heritage of mankind;

2. The area shall not be subject to appropriation by any means by States or persons, natural or juridical, and no State shall claim or exercise sovereignty or sovereign rights over any part thereof;

3. No State or person, natural or juridical, shall claim, exercise or acquire rights with respect to the area or its resources incompatible with the international regime to be established and the principles of this Declaration;

4. All activities regarding the exploration and exploitation of the resources of the area and other related activities shall be governed by the international regime to be established;

5. The area shall be open to use exclusively for peaceful purposes by all States whether coastal or land-locked, without discrimination, in accordance with the international regime to be established;

6. States shall act in the area in accordance with the applicable principles and rules of international law including the Charter of the United Nations and the Declaration on Principles of International Law concerning Friendly Relations and Co-operation among States in accordance with the Charter of the United Nations, adopted by the General Assembly on 24 October 1970, in the interests of maintaining international peace and security and promoting international co-operation and mutual understanding;

7. The exploration of the area and the exploitation of its re-

sources shall be carried out for the benefit of mankind as a whole, irrespective of the geographic location of States, whether land-locked or coastal, and taking into particular consideration the interests and needs of the developing countries;

8. The area shall be reserved exclusively for peaceful purposes, without prejudice to any measures which have been or may be agreed upon in the context of international negotiations undertaken in the field of disarmament and which may be applicable to a broader area.

One or more international agreements shall be concluded as soon as possible in order to implement effectively this principle and to constitute a step towards the exclusion of the sea-bed, the ocean floor and the subsoil thereof from the arms race;

9. On the basis of the principles of this Declaration, an international regime applying to the area and its resources and including appropriate international machinery to give effect to its provisions shall be established by an international treaty of a universal character, generally agreed upon. The regime shall, *inter alia*, provide for the orderly and safe development and rational management of the area and its resources and for expanding opportunities in the use thereof and ensure the equitable sharing by States in the benefits derived therefrom, taking into particular consideration the interests and needs of the developing countries, whether land-locked or coastal;

10. States shall promote international co-operation in scientific research exclusively for peaceful purposes:

(a) By participation in international programmes and by encouraging co-operation in scientific research by personnel of different countries;

(b) Through effective publication of research programmes and dissemination of the results of research through international channels;

(c) By co-operation in measures to strengthen research capabilities of developing countries, including the participation of their nationals in research programmes.

No such activity shall form the legal basis for any claims with respect to any part of the area or its resources;

11. With respect to activities in the area and acting in con-

formity with the international regime to be established, States shall take appropriate measures for and shall co-operate in the adoption and implementation of international rules, standards and procedures for, *inter alia*:

(a) Prevention of pollution and contamination, and other hazards to the marine environment, including the coastline, and of interference with the ecological balance of the marine environment;

(b) Protection and conservation of the natural resources of the area and prevention of damage to the flora and fauna of the marine environment;

12. In their activities in the area, including those relating to its resources, States shall pay due regard to the rights and legitimate interests of coastal States in the region of such activities, as well as of all other States, which may be affected by such activities. Consultations shall be maintained with the coastal States concerned with respect to activities relating to the exploration of the area and the exploitation of its resources with a view to avoiding infringement of such rights and interests;

13. Nothing herein shall affect:

(a) The legal status of the waters superjacent to the area or that of the air space above those waters;

(b) The rights of coastal States with respect to measures to prevent, mitigate or eliminate grave and imminent danger to their coastline or related interest from pollution or threat thereof resulting from, or from other hazardous occurrences caused by, any activities in the area, subject to the international regime to be established;

14. Every State shall have the responsibility to ensure that activities in the area, including those relating to its resources, whether undertaken by governmental agencies, or non-governmental entities or persons under its jurisdiction, or acting on its behalf, shall be carried out in conformity with the international regime to be established. The same responsibility applies to international organizations and their members for activities undertaken by such organizations or on their behalf.

Damage caused by such activities shall entail liability;

15. The parties to any dispute relating to activities in the area

and its resources shall resolve such dispute by the measures mentioned in Article 33 of the Charter of the United Nations and such procedures for settling disputes as may be agreed upon in the international regime to be established.

At the same time, the General Assembly adopted, with 108 votes in favor, 7 against (the Soviet block), and 6 absentees, a resolution deciding to convene in 1973, a conference on the laws of the seas, which would deal not only with the establishment of an "equitable international regime—including an international machinery—for the area and resources of the seabed and subsoil beyond the limits of national jurisdiction," but also with "a broad range of related issues on the laws of the seas." This includes, most importantly, the breadth of the territorial seas, on which the Latin American states hope to push their claims for vastly extended national jurisdiction, and to obtain the support of other states. Success in this move would indicate a strong trend towards regional seas, with exclusive rights for the states of the region. Soviet opposition to this second resolution is due not only to hostility to an international seabed authority but mainly to the vast investment and paramount interest of the Soviet Union in world-wide fishing. The Seabed Committee, enlarged from 42 to 86 members, is to prepare the relevant draft conventions in 1971.

While these resolutions constitute a modest victory for the forces opposing continuing partition of the oceanbed among coastal nations, the failure to determine the "limits of national jurisdiction" and the absence of a moratorium on further licenses is likely to lead to further expansions and national claims pending the uncertain conclusion of a treaty at some further date.

A Bibliographical
Note

TWO recent symposia give a comprehensive analysis
of the different aspects of seabed exploration, with contributions from oceanographers, marine biologists, strategic experts, economists, and lawyers of many countries. One is *The Symposium on the International
Regime of the Seabed*, published in Rome by the Accademia Nazionale dei Lincei in 1970 and referred to
in the present book as *Rome Symposium*. It is divided
into five parts:

PART   I   The Configuration of the Ocean Floor
           and Its Subsoil; Geopolitical Implica-
           tions

PART   II The Economic Resources of the Seabed

PART III The Present Regime and Possible Future
           Regimes of the Seabed

PART   IV The Military Uses of the Seabed and
           Their Regime

PART   V   Scientific Research on the Seabed and
           Its Regime

The other is a collection of five volumes, also of an

interdisciplinary character, scheduled for publication in its entirety in 1971 by Dodd, Mead and referred to in this volume as *Pacem in Maribus*. The five volumes—each of which contains from twelve to fifteen individual articles—are entitled as follows:

VOLUME I *Quiet Enjoyment: Arms and Police Forces for the Ocean*

VOLUME II *The Emerging Ocean Regime: Area of Competence and Legal Framework*

VOLUME III *Planning and Development in the Oceans*

VOLUME IV *Ocean Enterprises: A Summary of the Prospects, and Hazards, of Man's Impending Commercial Exploitation of the Underseas* (published in 1970 by the Center of Democratic Institutions)

VOLUME V *Ecology and the Role of Science and Scientists in an Ocean Regime*

The most important legal symposium is *The Law of the Sea: Proceedings of the Second Annual Conference of the Law of the Sea Institute*, ed. Louis M. Alexander (Kingston, R. I.: University of Rhode Island, 1967). The two leading legal monographs in the English language are J. Andrassy's *International Law and the Resources of the Sea* (New York: Columbia University Press, 1970) and Louis Henkin's *Law for the Sea's Mineral Resources* (New York: Institute for the Study of Science in Human Affairs, Columbia University, 1968). *Towards a Better Use of the Ocean* (1969), a monograph published by the Stockholm International Peace

Research Institute (SIPRI), consists of an analysis by Professor W. T. Burke, "Contemporary Legal Problems in Ocean Development," followed by Comments and Recommendations by an International Symposium held in June, 1968.

The most authoritative survey of oceanbed resources is John L. Mero's *The Mineral Resources of the Sea* (Elsevier, 1965).

A special issue of *Scientific American* (September, 1969) contains articles on the physical, oceanographic, biological, and ecological features of the ocean, as well as surveys of its mineral and food resources, by leading authorities.

A concise survey of the most recent developments in ocean mining and farming is given in *The Economist*, December 5, 1970, pp. 60–61.

INDEX

abyssal plain, defined, 12
adjacency criterion, 41–42
African Development Bank, 101
Air Afrique, 103
AGIP. *See* Ente Nazionale Idrocarburi (ENI)
American Petroleum Institute, 40, 73
Andrassy, Professor J., on sharing of wealth of land-mass prolongation, 42; on definition of continental shelf, 79
Asian Development Bank, 101
atolls, 12–13, 46–47

Bethlehem Steel Company, LAMCO joint venture, 104
Boyd-Orr, Sir John, 13
Brazil, "lobster war," 32. *See also* Latin America
British Gas Council, 94
British Petroleum, 97
Brown, E. D., on limits of continental shelf, 39

Canada, claim to 12-mile territorial zone and 100-mile "environmental protection" zone, 44–45, 109
Center for the Study of Democratic Institutions, proposal for international seabed authority, 89. *See also Pacem in Maribus*
Chile, copper exploitation, 22; seabed claims, 52. *See also* Latin America
China, participation in international seabed arrangements, 56, 83–84
Christy, Dr. Francis T., 15
Commission on Marine Science, Engineering and Resources, 22
Commission to Study the Organization of Peace, Report on The United Nations and the Bed of the Sea, 79, 88
Compagnie Française des Petroles (CFP), 97–98
COMSAT, 103
conservation of marine resources, 96, 107–111

contiguous zones, 31
continental margin, continental shelf, relation to, 39–42; defined, 12
continental rise, defined, 12
continental shelf, adjacency criterion, 41–42; continental margin, relation to, 39–42; customary law, relation to, 32–34; defined, 9–11; exploitability criterion, 12, 36–37, 39–41; freedom of the seas, infringement of, 2–3, 31–32, 34–35; graph, 10–11; islands, 37; research in, 59–60. *See also* landlocked states
continental slope, defined, 12
Convention for Seabed Exploration (Draft). *See* United States Draft Convention
Convention on Fishing and Conservation of the Living Resources of the High Seas, 108–109
Convention on the Continental Shelf, binding effect, 32, continental shelf, definition of, 12; drafting of, 35–38; exploration and research, 59; military installations, 56; revision of, 37; safeguards for fishing and shipping, 26, 38, 59
cooperation among nations, examples of, 115–117; prospects for, over seabed, 117–120
Council of Europe, 100
Cousteau, Jacques-Yves, 25
Craven, Dr. John P., 10, 49n., 54; on oceanbed transportation, 54–55

Declaration of Santiago, 43. *See also* Latin America
developing countries, development aid, need for, 92; joint ventures, advantages of participation in, 106–107, 112–113; mineral exporters, attitudes of, 22–23; position on international regime, 80–81; regional associations, role of, 101–102, 120. *See also* Latin America

International Seabed Resources Authority (U. S. Draft Convention), 72
International Whaling Convention, 107, 108
International Telecommunications Satellite Consortium (INTELSAT), 103
islands, continental shelves of, 37

Japan, pearl cultivation by, 32; overfishing by, 27, 108
Jennings, R. Y., on extent of continental shelf, 41–42
Johnson, Lyndon B., 28
joint ventures in ocean enterprise, developing countries, 106–107, 112–113; fishing, 110–112; government participation, 97–98; models, 99–107

Kellerman, Bernhard, 47
Kreditanstalt für Wiederaufbau, 104

LAMCO, 104
landlocked states, 14–15, 42–43, 80
LaQue, Dr. Frank, on estimates of revenue from seabed taxes, 78
Latin America, claims to 200-mile territorial zones, 15, 22, 43–44, 48n.9, 69, 81, 109; Declaration of Santiago, 43; nationalization of foreign mining interests, 80
Lauterpacht, Sir Hersch, on continental-shelf concept as customary international law, 34
licensing of ocean enterprise, principles of, 86–89; United States draft treaty, 73–75

Maltese proposals. See Pardo, Arvid
maritime powers, interests of, 3, 30–31; military advantages of, 51
Menard, Henry, 12–13
Middle East, political crises, effect on oil supply and claims, 80
military uses (seas and seabed), legal controls, 51–53; research in, 60; underwater transport, advantages of, 54–55. See also disarmament

mining (seabed), fishing, effect on, 26–27; navigation, effect on, 26; techniques, 23–26, 29n.7, 62
Mouton, Admiral M. W., 36
Munch, F., 40

national claims to seas and seabed, claims to superjacent waters, 35, 44–45, 50; exploitability criterion, 36–37, 39–41; inequalities between states, 14–16; population and technological pressures, 25–26, 28
National Iranian Oil Corporation (NIOC). See IMINOCO
National Petroleum Council, 63–64
nationalism, generally, 8; barrier to international cooperation, 117–118; increase in national states, 14; inequalities between states, increase of, 14–16; international regime, opposition to, 63–70, 79–80
navigation, capability of ships, 55; continental shelf, effect of, 2–3, 31–32, 34–35; oceanbed mining, effect of, 26; protection of, 38; "surface system," limitations of, 54; underwater transport, advantages of, 54–55
Nixon, Richard M., on United States oceans policy, 45, 70–73
North Sea Continental Shelf Case (International Court of Justice), 32–34, 41–42, 99–100
Norway, 27, 111

Ocean Enterprise, generally, 95, 112–113; regional banks, role of, 101–102; regulation of, 95–96. See also joint ventures; licensing
Oda, Professor Shigeru, on "exploitability" test, 40
Operations Committee (U.S. Draft Convention), 91
Outer Space Treaty, 57

Pacem in Maribus, 89, 122
Pardo, Arvid, survey of seabed mineral resources, 20; United Nations draft

resolution on deep-sea exploitation and international authority, *64 et seq.*, 92

Pearson Commission, Report on "Partnership in Development" (1969), 91

Pell, Claiborne, draft seabed treaty, 89

Phillips. *See* IMINOCO

Piccard, Albert, 25

pollution of the seas, 27, 67; Canadian measures, 44–45

Pontecorvo, G., on exploitation technology, 23

resources of the sea and seabed, generally, 17–18; deep seabed, 46–47, 78; marine food, 18–19, 23, 110; minerals, 19–24; scientific research, 58–59. *See also* conservation of resources

Rules and Recommended Practices Commission (U.S. Draft Convention), 90

safety zones, 56

Scelle, Professor Georges, on continental-shelf doctrine, 35

seabed, demilitarization of, 55–58; draft treaty on military use of, 53–54; legal status of, 46–48, 51–52, 53; scientific research, 58–59

Sealab, 25

seamounts, 12–13, 46–47

Selden, John, doctrine of *Mare Clausum*, 3, 30, 44

shallow banks, 12–13, 46–47

Shell Oil, 97

Soviet Union, attitude on military use of seabed, 52–55; fishing fleets of, 27, 108, 109; territorial waters claims, 31

Stewart, C., on exploitation technology, 23

Strategic Systems Project Office (U.S. Navy), 54

territorial waters, claims to 200-mile zones, 43–44; extension of, 31

Truman Proclamation, generally, 2, 3–6, 19, 31–32; definition of continental shelf, 6, 10, 72; text of, 6–7

underwater transport, 54–55

United Nations, Disarmament Committee, 52–53; failure to create system of international security, 83, 118; General Assembly, actions on seabed, 66–70; Pardo Resolution, 64–65, 92; Resolution on Permanent Sovereignty over National Resources, 111; resolutions on seabed and sea law, 121–125; Seabed Committee, 66–69, 73; Secretary General's "Study on International Machinery," 84–88; specialized agencies of, 115. *See also* United States Draft Convention

United States, claims to 9-mile "contiguous zone," 31; Endangered Species Act, 108; House Subcommittee report on seabed, 63; internal differences, 71, 80, 91; Nixon policy statement on oceans policy, 70–71; oceanbed activities, regulation of, 98; offshore exploitation, 94–95; proposal on military use of seabed, 52–55; Sealab experiments, 25; whale-oil imports, position on, 108. *See also* Truman Proclamation; United States Draft Convention

United States Draft Convention, conservation provisions, 96; delimitation of national jurisdiction, 72–73, 78–79; international trusteeship area, 73–75, 113; licensing of ocean enterprise, 73–74, 86; structure of international seabed authority, 72, 89–92; vested interests, protection of, 63, 72–73

United States Export-Import Bank, 104

United States Public Health Service, on sea farming, 110

Western European Union, 100

World Bank. *See* International Bank for Reconstruction and Development